SpringerBriefs i

For further volumes:
http://www.springer.com/series/10047

Syed Ather Hussain Akbari

Immigrants in Regional Labour Markets of Host Nations

Some Evidence from Atlantic Canada

Springer

Syed Ather Hussain Akbari
Department of Economics
St. Mary's University
Halifax, NS
Canada

ISSN 2211-3215 ISSN 2211-3223 (electronic)
ISBN 978-94-007-6243-5 ISBN 978-94-007-6244-2 (eBook)
DOI 10.1007/978-94-007-6244-2
Springer Dordrecht Heidelberg New York London

Library of Congress Control Number: 2012956376

Springer is part of Springer Science+Business Media (www.springer.com)

Foreword

This is a book for all those interested in how to improve the economic integration of immigrants. The book uses Atlantic Canada as a laboratory to see how immigrants fare in regional labour markets and to identify best policy practices that can be used by other countries to address demographic challenges similar to those facing Canada—such as population ageing and youth out-migration from smaller regions to larger regions—through immigration. An overarching result is that analysing regional data can lead to very different conclusions than analysing national data, which means that it can be risky to devise immigration policy based only on national data.

A key strength of the book is the extent of the statistics presented. Indeed, I would dare to say that I have never come across such a comprehensive and self-contained compendium of facts and statistics on the labour market performance of immigrants in Atlantic Canada. This book should not only provide an invaluable source of information to policy makers interested in the labour market performance of immigrants in Atlantic Canada, but also provide a research template for researchers and policy makers in other countries who are also grasping with the type of immigration issues facing Canada.

This book is also a trove of facts that call for further research. A case in point is the observation that immigrants in Atlantic Canada earn on average more than Canadian born individuals. This is intriguing since it is well known that for the whole of Canada, it is the other way around: immigrants earn less on average than Canadian born individuals despite being more educated and having more work experience. Such issue definitely deserves further research. Indeed it would probably be very instructive to know from a policy point of view why the labour market performance of immigrants in Atlantic Canada is so much better than that of immigrants in the rest of Canada.

Another interesting contribution of the book is the discussion on how international students contribute to immigration in Atlantic Canada and what have been the policy initiatives by provincial governments to attract and retain international students.

Finally, a few words about the author. Professor Akbari is the leading researcher on immigration in Atlantic Canada. It is therefore most fitting that he be the author of such a book.

October 1, 2012

Serge Nadeau
Director
Research Group on the Economics
of Immigration
University of Ottawa
Canada

Acknowledgments

This publication derives from several research projects I conducted for Atlantic Canada Opportunities Agency (ACOA, Government of Canada) over the period 2005–2008. The data in this publication have been updated until 2010 and some more recent immigration policy developments in Atlantic Canada have also been noted. These were recent at the time of writing of this manuscript. As well, I have also included some (albeit short) discussion of the impact of regionalisation of immigration in major host nations of the west.

I have benefitted from my discussions with ACOA officials, with the officials of the four provincial governments of Atlantic Canada, settlement organizations and with my colleagues around the world who conduct research on social and economic impacts of immigration. I also acknowledge the comments I received from an anonymous reviewer on an earlier version of the present manuscript.

Contents

Chapter 1
Introduction

Declining natural growth rates in population have resulted in shortages of skilled workers in all western countries. However, within each country, smaller regions are affected even more because of the added phenomenon of youth out-migration to larger regions. Immigration is seen as an important tool to reverse these trends in regional population growth rates. However, international migrants tend to settle in the large urban centres of their host countries, mainly because of greater economic opportunities and the presence of large social networks formed by the already established immigrant communities in these larger centres. Since the middle of 1990s, immigration policy in these countries is responding to the challenges posed by regional population imbalances through federal-regional collaborations that introduce special programs to attract immigrants to smaller areas and then retain them there. The main focus of regional/provincial and state policies in western countries to attract and retain immigrants has been the skilled immigrant.[1] The following brief overview of such policy initiatives undertaken in four major immigrant-receiving countries will provide the readers of this book some broad perspective on such efforts.

[1] A report published by Organization for Economic Cooperation and Development (OECD) also identifies the link between regional development policies and immigration in host countries of the west (OECD 2004).

S. A. H. Akbari, *Immigrants in Regional Labour Markets of Host Nations*, SpringerBriefs in Population Studies, DOI: 10.1007/978-94-007-6244-2_1,

Some Immigration Policy Initiatives Towards Regionalization in Australia, Canada, Germany and New Zealand

Australia

About 158,000 immigrants arrived in Australia during 2008–2009. More than 60 % of new arrivals in Australia settle in the states of New South Wales and Victoria, while the rest are divided among the remaining six states and the Capital Territory. The 2001 census found that about 85 % of immigrants lived in metropolitan areas of which 60 % residing in Sydney and Melbourne.[2] In 1995, Australia introduced state specific and regional migration (SSRM) visa schemes, which enhanced the role of state and regional authorities in Australia's immigration program. These schemes explicitly integrate international migration with regional development planning and strategies by enabling[3] state and territorial governments and regional employers to influence the number and profile of skilled migrants settling in their areas in line with both the regional demand for skills and their development objectives. In fact, as Hugo (2008) notes, two classes of immigrants were created under these schemes. Immigrants under one class were free to choose wherever they wanted to settle in Australia, while those under the other class were restricted in where they could settle, at least during their initial years (normally three) in Australia. In 2005, Australia attracted about 16 % of its new arrivals under its regional migration scheme, by 2012 it attracted about a quarter of new arrivals under this scheme.

Canada

Canada receives about 250,000 immigrants each year. More than 40 % settle in the province of Ontario, while about 45 % settle in Alberta, British Columbia and Quebec combined. The remaining 15 % are thinly divided among the seven smaller provinces and territories. Canada's Provincial Nominee Program (PNP) is implemented along the same lines as Australia's SSRM visa schemes. Under the PNP, a province or a territory nominates an individual for immigration to Canada who will live in that same province or territory. The relevant individual has the skills, education and work experience needed to make an immediate economic contribution to the nominating province or territory. The PNP agreement was first

---

[2] http://www.workpermit.com/news/2006_01_30/australia/rural_areas_attracting_immigrants.htm

[3] The source of 2001 data is the same as listed in the above footnote. The 2012 data source is Australian Government (2012).

signed in 1991 between the federal government and the province of Quebec, which also has greater jurisdiction over its immigration program than do other provinces. The agreement with Manitoba, first signed in 1996, had the specific objective of meeting a skilled labour shortage in that province. Since then, other provinces have also signed PNP agreements to meet their labour market needs. Changes introduced at the federal level also allow international students at Canadian post-secondary institutions to work in Canada for up to 2 years after graduation provided they work outside of Montreal, Toronto and Vancouver, the major population centres and destinations of most immigrants. This arrangement also gives them a greater potential to successfully apply under a PNP.

Germany

Germany recognized the importance of regional labour market needs in the German Immigration Act, which went into force in 2005. In general, two independent authorities, i.e., the Aliens Authority and the Federal Employment Agency (FEA), assess an application for a residency permit filed for employment purposes. The Aliens Authority acts at the community level to examine whether the application meets the general rules set by policy with respect to immigration, integration and social suitability. The FEA examines the suitability of the application in light of local labour market needs.[4] Germany received 603,314 foreigners in 2009: about 48 % of them came from Europe, and about 71 % of these Europeans were from Eastern Europe. Most of the remaining 52 % arrived from Turkey, China, the United States, the Russian Federation, Iraq and India.

New Zealand

Based on the Government of New Zealand (Ministry of Business, Innovation and Employment) data, New Zealand received an average of about 84,000 permanent and long-term arrivals each year during 2001–2012.[5] Maré et al. (2007) reported that at the time of the 2001 census, about 60 % of immigrants resided in South Auckland, 10 % in Wellington and 8 % in Christchurch. These three large cities provide strong labour market opportunities. Unlike in Australia and Canada, the absence of a state or provincial authority in New Zealand has meant that policies to both promote immigration and meet labour market needs have been centrally driven. Local labour market considerations have dominated New Zealand's

---

[4] For further details, see Burkert et al. (2008).

[5] http://www.dol.govt.nz/research/migration/monthly-migration-trends/12jun/data.asp?id=fig1

immigration policy in this century. Spoonley and Bedford (2008) report increasing regional migration initiatives adopted as a result of a collaboration among local agencies, employers, city councils and the New Zealand Education Commission. These authors cite two examples: one from Southland, a region dominated by primary production, tourism and small towns, and the other from Waitakere, the largest urban agglomeration, whose economy is dominated by small- and medium-sized businesses. In the case of Southland, Venture Southland, an economic development agency, has played an active role in collaboration with the Southland Chamber of Commerce and Engineering South in recruiting immigrants from the UK to its agriculture and tourism industries. The Waitakere case is an example of similar cooperation among various stakeholders, such as the Waitakere City Council, Skills New Zealand/Tertiary Education Commission and Enterprise Waitakere. These groups identified significant skill shortages in small- and medium-sized enterprises. The collaboration works by facilitating the provision of information to both employers to recruit immigrant workers and to immigrant workers about local labour market requirements.

The Impact of Immigrant Regionalization on the Geographic Distribution of Immigrants in Canada

While regional policy initiatives in all four countries have been discussed in the literature, direct statistical evidence on the impact of these changes on the geographic distribution of immigrants can only be found for Canada.[6] Alberta, British Columbia, Ontario and Quebec are the major destinations of immigrants in Canada, receiving about 88 % of all immigrants in 2010, with Ontario alone receiving about 42 % that year. However, as Chart 1 shows, the immigrant arrival rate in Ontario went down from about 13 per thousand residents in 2001 to less than 8 per thousand in 2010. All other provinces experienced an increase over this same period. Increases in smaller provinces such as Manitoba, Saskatchewan and Prince Edward Island were significant. Further, all provinces in Atlantic Canada, which have traditionally received fewer immigrants than the rest of Canada, are now attracting more immigrants than before.

[6] Changes in regional distribution have also been observed in other countries, but it is not clear whether they were a result of immigrant regionalization policy aimed at meeting skill shortages in local and regional markets. Thus, for Australia, Hugo and Harris (2011) have observed a shift in the distribution of immigrant population since the mid-1990s away from Sydney (which still remains the major destination) towards other cities such as Victoria. However, Galligan et al. (2011) attributes these changes to not only policies aimed at meeting local skill shortages but also to a concentration in gateway cities. Data obtained on immigrant population distribution in Germany by this author also suggest decreases in arrival rates for all areas in Germany, except for Berlin and Bremen. However, it is not clear whether this was a result of any immigration policy changes (The author thanks Carola Burkert for her help with the data and for providing information on Germany).

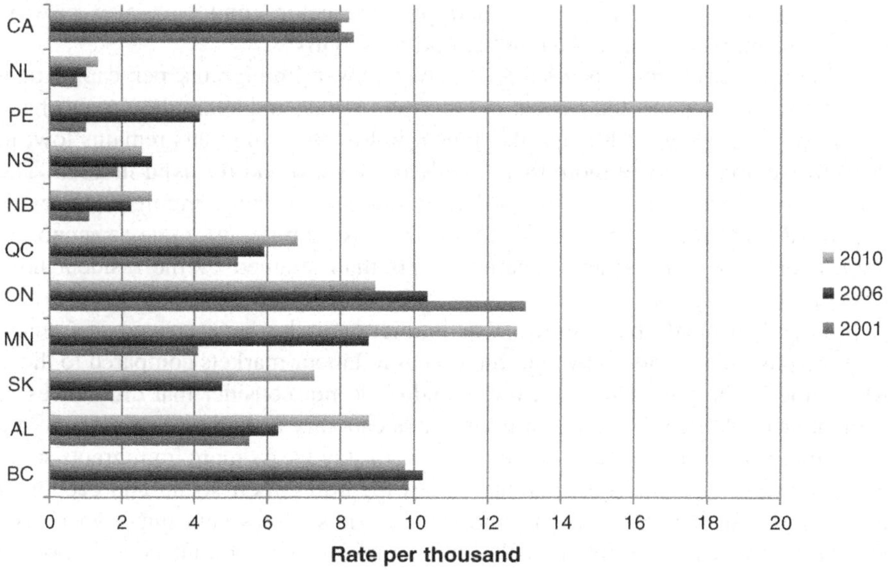

Chart 1 Immigrant arrival rates per thousand persons, Canada and Provinces. *Source* Table A1

About This Book

This book is about the outcomes of the regionalization of immigration in western countries with special focus on economic outcomes. The book's focus is on Atlantic Canada, a region located on the Atlantic coast of Canada comprising four provinces: Newfoundland and Labrador, New Brunswick, Nova Scotia, and Prince Edward Island. As the next chapter discusses in detail, with a total population of about 2.3 million, these four provinces share many demographic characteristics. For example, all have low fertility rates, which have continued to decline in the post-World War II period. Out-migration rates have been among the highest in the country. In fact, the population growth rate in these provinces has been among the lowest in Canada. The region's population also has the highest percentage of seniors, those aged 65 and over, in the country.

In response to the above demographic trends, which have begun to have negative economic consequences for the region, the four provinces in Atlantic Canada have adopted unified policies on many fronts. Immigration is considered an important policy tool to reverse the negative impacts of declining population growth. This emphasis on immigration is evident in the formation of separate ministerial departments with the mandate to raise immigration levels and their retention rates. While immigration selection remains under federal jurisdiction, each province plays a greater role in the national immigration program to attract immigrants based on its labour market needs. Municipal governments, community organizations, settlement agencies and private sector employers also provide inputs into immigrant attraction and retention initiatives.

Atlantic Canada has historically received fewer immigrants per capita than other Canadian regions, but current initiatives to attract and retain immigrants have begun to show results. Although the annual inflow of immigrants remains low, in 2010 the region received more than double the level it had received in 2000. The smallest province of Prince Edward Island saw its per capita immigration levels rise by 18 times during this period (Chart 1). Most immigrants come as economic immigrants and with higher education levels than attained by the resident non-immigrant population.

Studies in Canada have shown that immigrants who have arrived in Canada over the past two decades have not fared well in labour markets compared to those who came in the past. However, these studies do not consider that differences in the labour market performance of immigrants can vary according to the economic and social conditions of the region they chose to stay in. As more immigrants settle in smaller areas of their host nations, issues regarding their social and economic integration and their economic performance in those areas gain importance. The factual information presented in this book makes an important contribution to literature on immigration research by analyzing regional data on immigrants' economic performance. It presents strong evidence that immigrants in Atlantic Canada's labour markets do better than do resident non-immigrants. This finding is in sharp contrast with the findings of some other studies in Canadian literature which analyze national data and arrive at the conclusion that overall, immigrants perform poorly compared to native-born in Canadian labour markets.[7]

The study also analyzes the trends in the inflow of international students, who are viewed as potential skilled immigrants as they acquire relevant professional training and develop their knowledge of local language, culture and institutions. They are also sources of revenue to local educational institutions and help strengthen local markets for goods and services. Canada receives over 150,000 international students each year. Recent attempts by post-secondary institutions in Atlantic Canada to attract international students have resulted in their increased inflow to the region, which has, in turn, benefited the economy of Atlantic Canada.

The factual discussion in this book presents Atlantic Canada as a model of best practices for smaller regions of other western countries faced with similar demographic challenges and that are trying to use immigration as a tool to meet them. The evidence presented shows that immigration policy can work in attracting and retaining immigrants in smaller areas where economic opportunities exist for new arrivals, despite the assertion in the immigration literature that immigrants tend to settle in areas that have a large resident community from their countries of origin. In fact, Maré et al. (2007) also found that immigrants in New Zealand choose their location based on labour market opportunities rather than on the presence of migrant networks. Better labour market performance of

[7] A review of this literature can be found in Nadeau (2011).

immigrants in Atlantic Canada also implies that economic returns from immigration policy can be enhanced by changing the geographic distribution of immigrants within the country.

The main strength of the analysis presented in this book is its use of immigration data for a small area of Canada for a 30-year period based on various databases of Statistics Canada and Citizenship and Immigration Canada (CIC) including census files, administrative tax files, labour force survey data as well as some limited interviews and anecdotal evidence. The work presented examines a range of labour market outcomes that, where possible, distinguishes between recent and all immigrants, and disaggregates by country of birth. Yet another strong aspect of this work is the use of extensive charts and tables that summarize the main findings. Such an analysis has not yet been conducted for any other host country of immigrants. The author is confident that the information contained in this book will be useful for policy makers, immigrant settlement organizations, and academic researchers who may wish to pursue some small area immigration-related issues in greater depth. Finally, the book also highlights a need for consistent data collection on immigrant settlement and their labour market performance in small areas of host nations so the success and impact of immigrant regionalization can be assessed on a continuous basis.

References

Australian Government (2012). 2011–2012 Migration Program Report Program year to 30 June 2012. Department of Immigration and Citizenship.

Burkert, C., Niebuhr, A., & Wapler, R. (2008). Regional disparities in employment of high-skilled foreigners—determinants and options of labour market policy in Germany. *Journal of International Migration and Integration, 9*(4), 383–400.

Galligan, B., Boese, M., Phillips, M., & Kearton, A. (2011). *Boosting regional settlement of migrants and refugees in Australia: policy initiatives and challenges*. Melbourne, Australia: School of Social and Political Sciences, The University of Melbourne.

Hugo, G. (2008). Australia's state specific and regional migration scheme: an assessment of its impacts in South Australia. *Journal of International Migration and Integration, 9*(2), 125–145.

Hugo, G.J., & Harris, K. (2011). *Population distribution effects of migration in Australia*. Condensed Version. Report for Department of Immigration and Citizenship. Australia:University of Adelaide.

Maré, D.C., Stillman, S., & Morten, M. (2007, September). *Settlement patterns and the geographic mobility of recent migrants to New Zealand*. Motu Working Paper 07-11 Motu Economic and Public Policy Research.

Nadeau, S. (2011). The Economic Contribution of Immigration in Canada—Recent Developments: What do we know? What does it mean for policy? Mimeo, Research Group on the Economics of Immigration, University of Ottawa.

OECD. (2004). Trends in international migration: SOPEMI 2003 edition (ISBN 92-61-01944-8).

Spoonley, P., & Bedford, R. (2008). Responding to regional labour demand: international migration and labour markets in New Zealand's regions. *Journal of International Migration and Integration, 9*(2), 203–223.

Chapter 2
Some Demographic Trends in Atlantic Canada: Potential Consquences and Policy Response

The population growth rate in Atlantic Canada has been in continuous decline for most of the post-World War II period, the most drastic decline taking place after the 1970s. Net out-migration from the region has been a permanent factor in this decline. This means that to maintain some population growth, the region has relied solely on natural increase (births minus deaths). However, this component of population growth has also declined continuously, falling below even the out-migration rate in the new millennium, thereby causing the population *growth rate* to become negative in the first half of the last decade. During 2006–2010, the population growth rate rose but remained very close to zero (i.e., an increase of only about 14,350 in the region) because of a lower net out-migration rate than occurred at the beginning of the century and one observed largely due to an increase in international immigration during this period. Chart 1 shows the trends in components of population growth.

Chart 2 shows that since 1945, the declining natural population growth rate in Atlantic Canada was due mainly to declining birth rates because death rates remained almost constant over most of the period. Since 2003, the gap between birth and death rates has been below one per thousand annually, and the region's population reached a zero natural growth rate in 2010.

Table 1 shows more detailed components of population growth in each Atlantic province during 2009–2010, during which the region's population increased by about 11,000. About 62 percent of this increase was generated by international migration; only 1.5 percent came from a net natural increase. The contribution of international migration towards population growth was the least in Newfoundland and Labrador, where net *interprovincial* migration contributed the most (62 percent), largely due to the continuing development of offshore oil reserves. Immigration contributed the most to the population growth of Prince Edward Island, where net interprovincial migration was negative; in fact, the province actually would have suffered a population decline without immigration. Natural growth was also negative in the populations of Nova Scotia and of Newfoundland and Labrador during 2009–2010.

S. A. H. Akbari, *Immigrants in Regional Labour Markets of Host Nations*,
SpringerBriefs in Population Studies, DOI: 10.1007/978-94-007-6244-2_2,

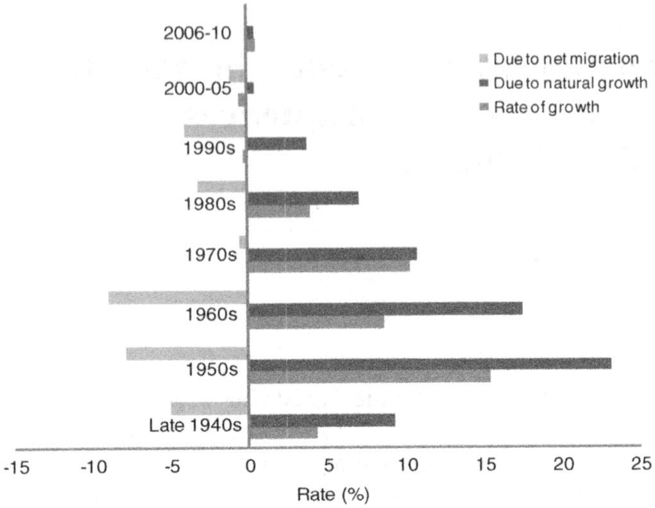

Chart 1 Components of population growth rates in Atlantic Canada in the post-World War II period. *Source and notes* Table A2. A negative growth rate means population declined during that period

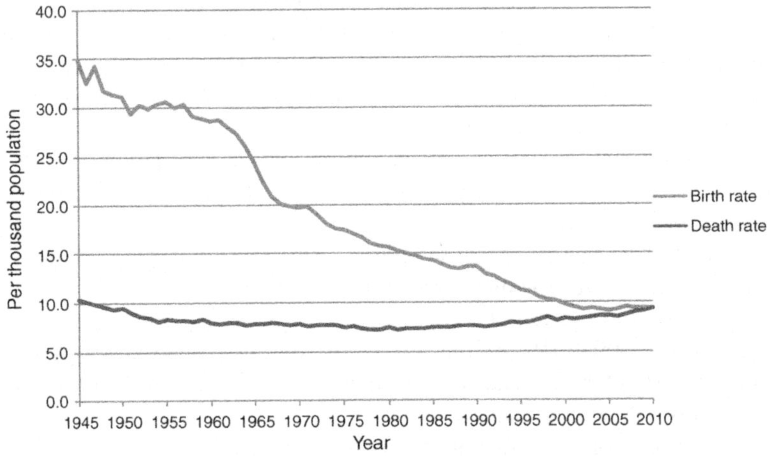

Chart 2 Birth and death rates, Atlantic Canada, 1945–2010. *Source* Table A3

Despite the small share of international immigrants that Atlantic Canada receives nationally, their contribution to population growth has been strong since the 1980s, as shown by Akbari (2009), based on data from each of the six censuses conducted during 1981–2006. During the 1990s, population growth would have been negative without international immigration in the region, while during 2001–2006, immigrants accounted for more than half of population growth in the region.

Table 1 Components of population growth, Atlantic Provinces, 2009–2010

	July 1, 2009/June 30, 2010				
	NL	PE	NS	NB	Total
Births (1)	4,480	1,397	8,810	7,044	21,731
Deaths (2)	4,765	1,217	8,840	6,743	21,565
Immigration (3)	659	1,791	2,409	1,928	6,787
Emigration (4)	372	77	714	527	1,690
Net temporary emigration (5)	258	70	474	380	1,182
Returning emigrants (6)	120	71	442	338	971
Net non-permanent residents (7)	423	150	1,544	49	2,166
Net interprovincial migration (8)	1,309	−876	205	722	1,360
Population growth (9) = (1) − (2) + (3) − (4) − (5) + (6) + (7) + (8)	2,112	1,309	4,330	3,191	10,942
Immigrants in population growth (% = (3) of (9)	31	137	56	60	62

Source Statistics Canada, CANSIM, Table 051-0004 (Last modified: 2010–09–29)

The slowing of Atlantic Canada's natural increase in population has resulted in its having the highest percentage of seniors among all Canadian regions. Statistics Canada (2011) reports that as of July 2011, Nova Scotia had the highest percentage of seniors in Canada (16.5 percent), followed by New Brunswick (16.2 percent), while the provinces of Newfoundland and Labrador and Prince Edward Island had identical percentages (15.8 percent). Newfoundland and Labrador had the highest median age (43.8 years). In terms of youth, Nova Scotia had the lowest percentage in the country (14.7 percent), very closely matched by Newfoundland and Labrador (14.8 percent). Nova Scotia and New Brunswick were among the four Canadian provinces where the percentage of youth in the population was below that of seniors, the other two being British Columbia and Quebec. The median age in all Atlantic provinces exceeded that everywhere else in the country.

Potential Consequences of Population Decline and Aging

Population decline and aging in a region can have several economic and political consequences, some of which are discussed below.

Some Economic Consequences

Population decline can result in shrinking markets for goods and services, thereby creating an adverse impact on incentives for business investment. One might expect the population decline to result in the creation of policies for restructuring the economy to provide the goods and services demanded by a growing elderly population. However, such restructuring could not have a long-lasting effect because the death rate is soon expected to exceed the birth rate in Atlantic Canada (Chart 2). A decline of some federal fund transfers to the region, such as federal funds for health care, is also expected because these are determined by a province's population.

An aging population can also result in fewer labour force participants, thereby causing shortages of certain types of labour demanded by employers in the region. For example, Skills Compétences Canada (date unknown) identified shortages of construction workers, especially bricklayers and electricians, in Atlantic Canada. Shortages of professionals such as health care providers in the region are also well documented. A study by the Atlantic Institute for Market Studies (McNiven and Foster 2009) predicted that by 2016, the number of available workers in Nova Scotia will be lower than the number of available jobs and that by 2026, approximately 12.5 percent of jobs will be vacant in the province. A similar study conducted by the Policy Research Centre at the University of New Brunswick, and reported by the Atlantic Institute for Market Studies (2009), predicted that New Brunswick's labour force would begin to decline as early as 2011. In 2007, the Newfoundland and Labrador Skills Task Force released a report (Newfoundland and Labrador 2007a) that predicted serious skilled labour shortages for many of that province's large-scale development projects in the coming years. Further, according to the Canadian Tourism Research Institute and Conference Board of Canada (2010), Prince Edward Island's labour force is expected to grow between 2010 and 2030 but by a mere 0.2 percent. The tourism industry alone is expected to have 1,200 unfilled jobs by 2025 in Prince Edward Island.

Rural Atlantic Canada has been affected the most by regional population decline, as noted in a study published by the Atlantic Provinces Economic Council (2007b). This phenomenon is of concern because most natural resource-based industries in Atlantic Canada are located in rural areas. Population decline would cause shortages of skilled labour in these industries. In addition, when populations decline, the cost of public, as well as private, services does not adjust immediately, and there is a point below which base costs cannot go regardless of population size. As a result, the economic feasibility of providing such services becomes questionable. Losses of hospitals and mail and banking services, as well as consolidation of schools in rural Atlantic Canada, have become increasingly common. This closure of public and private services then further accelerates rural population decline as people move closer to metropolitan areas in search of those services.

Last, but not the least, an aging population can also increase pressure on younger labour force participants to provide the social programs for the elderly (for example, higher contributions to government pension plans and higher taxes).

1 Nova Scotia's Aging Workforce

"Nova Scotia's labour force is strongly affected by changing demographics; the aging population will lead to increased retirements in the coming years. Workers over the age of 45 now represent about 51 % of Nova Scotia's labour force, while for Canada overall, the rate is 48 %, suggesting Nova Scotia has an older population than the average Canadian province. Current projections of the decline in the size of the labour force associated with age suggest a 20 % decrease in the labour supply. This could mean that Nova Scotia will face a situation in which the number of jobs will exceed the number of people willing to participate in the labour market. The period between 2006 and 2016 is seen as one of transition from an era of labour surpluses to one of chronic shortages". Ray Barton Associates Ltd. (2008, p. 13).

Some Political Consequences

Population decline can also result in a weakening of a province's political representation in the Canadian House of Commons.[1] According to the *Constitution Act* of 1985, each province's representation is based on its population size, although a grandfather clause protects each province from losing seats in the House below its 1976 level. This method of determining a province's number of seats in the House implies that as the population of a province grows, not only would the number of seats allocated to it increase but also its total number of seats in the House above its 1976 level. In turn, proportionate representation of other provinces whose population either remains stable or falls would decline.

A study published by the C. D. Howe Institute (Tomlin 2007) noted the imbalance in provincial representation in the House of Commons resulting from population imbalance among the provinces. The study also projects that if these current uneven trends in population growth rates across provinces continue, the imbalance in provincial representations will worsen by 2021. As Chart 3 shows, each Atlantic province had less than 4 percent of the total seats in the House of Commons in 1976. By 2021, declining population is projected to further weaken this representation. Most recently (as of October 27, 2011), the federal Conservative government introduced a Fair Representation Bill in the parliament to add 30 new seats to the House of Commons for Alberta (6), British Columbia (6) Ontario (15) and Quebec (3), while all other provinces will maintain their current number. These additions to the House seats are to take place from the time of next expected election in 2015 which will reduce the Atlantic provinces' percentage

[1] This can also have an indirect economic effect on the region.

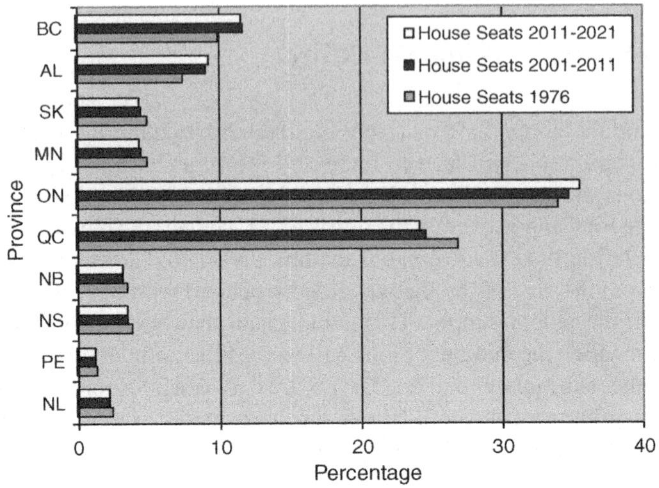

Chart 3 Provincial representation in the Canadian house of commons based on current rules and projected population growth rates. *Source* Table A4

representation in the House as follows: Newfoundland and Labrador down from 2.09 to 1.41 percent, Prince Edward Island down from 1.41 to 0.42 percent, Nova Scotia down from 3.61 to 3.28 percent, and New Brunswick down from 3.28 to 2.99 percent.

Public Policy and Community Responses to Population Decline and Aging

The negative economic and political consequences of population decline and aging have not gone unnoticed by policy makers in Atlantic Canada. Therefore, to help reverse its negative demographic trends, each province has launched a population strategy, of which increasing the provincial share of annual Canadian immigrant inflows is an important component. Separate government departments have been formed with the mandate to increase (1) the level of skilled worker immigration as one way to deal with skill shortages, and (2) the retention rate of annual immigrant inflows, which has been low in most Atlantic provinces since the mid-1990s. Each government is now (1) a signatory to a Provincial Nominee Program (PNP), a federal-provincial bilateral agreement that allows each participating Canadian province to target and recruit immigrants to meet its own particular needs and who are then fast-tracked through the system by Citizenship and Immigration Canada (CIC), and (2) collaborating with other provincial governments in the region and federal departments to develop promotional material, participate in overseas marketing missions, conduct research, assess and recognize foreign credentials and share information.

The Atlantic Canada Opportunities Agency (ACOA), a federal government agency with a regional economic development mandate, also recognizes the importance of immigration to the economic development of the region. ACOA works closely with both the region's governments and its settlement organizations to facilitate the settlement and integration of immigrants. ACOA established an Atlantic Population Table (APT) with representation from CIC, Human Resources and Social Development (HRSD), and the four Atlantic provincial governments. Another federal government department, the Rural Secretariat, has investigated a rural re-population strategy, with immigrant settlement in rural regions as one component of rural economic development. Further, the Nova Scotia Department of Agriculture has partnered with the Nova Scotia Office of Immigration to attract immigrant farmers to rural areas of the province.

Municipal governments and the private sector also realize both the importance of immigration in meeting local labour market demands and the need to build welcoming neighbourhoods to retain immigrants. For that reason, immigrant attraction and retention are listed as key priorities in the economic strategy document of Halifax Regional Municipality (HRM).[2] HRM, in collaboration with the Greater Halifax Partnership (GHP), has launched an Immigration Action Plan (IAP) to promote economic and cultural development in HRM,[3] which is an example of public–private sector collaboration to meet local labour shortages in the region.

Settlement organizations such as the Immigrant Settlement & Integration Services in Nova Scotia (ISISNS), the PEI Association of Newcomers in Prince Edward Island (PEIANC), and the Association for New Canadians in Newfoundland and Labrador (ANCNL), along with community organizations such as the Colchester Regional Development Authority (CoRDA) in Nova Scotia and the Multicultural Association of Carleton County in New Brunswick, work closely with their provincial and municipal governments to design innovative strategies to attract immigrants and facilitate their integration. Consultations are held with newcomers to help them find a job and establish a business. Among other services, ISISNS also provides pre-arrival services to those approved to settle in Nova Scotia, while ANCNL provides a pre-employment readiness service and also participates in the Acquiring eXperience and Integrating Skills (AXIS) career service (see details in Box 2.2).

The Council of Atlantic Premiers (CAP) has established a unified goal to increase international immigration in the Atlantic region.[4] Recognizing the importance of regional collaboration in attracting and retaining skilled immigrants

---

[2] This document can be found at http://www.halifax.ca/economicstrategy/EconomicStrategy.html.

[3] The GHP is a public–private sector coalition aimed at attracting new investments in Halifax.

[4] The Atlantic provinces have adopted unified policies on many fronts and cooperate in providing many public services. For example, the sales tax is harmonized across all provinces, which also cooperate in providing selected health care and education services. Businesses have also developed linkages across the region, especially in professional services. Proposals for creating a single economy in the region have also been tabled (Atlantic Provinces Economic Council 2007a).

to meet the region's workforce needs, CAP had been working since 2005, with the federal government through the Atlantic Population Table (APT) to achieve this purpose as one of its larger goal of developing the region's labor force. The funding agreement supporting the APT expired in January 2012, but the four provincial governments have now established an Atlantic Workforce Partnership with the objective of developing the region's workforce to meet the changing skill requirements especially in mining, energy and shipbuilding capital projects. Immigration remains one of the top priorities of this partnership among Atlantic provinces.[5]

2 Employment Assistance to New Immigrants in Newfoundland and Labrador

The Association of New Canadians in Newfoundland and Labrador provides several employment services to new immigrants in the province. One is the AXIS (Acquiring eXperience; Integrating Skills) service, which provides comprehensive career-focused services and employment programs for educated professionals, trades persons and entrepreneurs. Through interactive engagement with employers and collaborative partnerships with all levels of government, industry and the community, AXIS aims to facilitate successful labour market integration for newcomers.

Following an initial assessment, clients and counsellors develop an individualized Career Action Plan, which often includes facilitation of credential recognition and participation in pre-employment readiness training. Through the incorporation of one-on-one counselling and group seminars, clients receive comprehensive and current information in preparation to enter the Canadian workplace. Clients also have the opportunity to participate in customized Bridge-to-Work programs, including Mentoring Link, STEP Career Training Placements, the Internship Placement Program (IPP), and Entrepreneurial Assistance.

The Skills-Matching Database is an online recruitment tool where employers can post job openings for free, access online client résumés, find answers to immigration questions and labour market information, and access services available through AXIS Career Services. AXIScareers.net quickly and effectively links a talented pool of work-ready, internationally educated and skilled workers with employers in the province of Newfoundland and Labrador.

Source www.ancnl.ca

[5] http://www.gov.ns.ca/news/details.asp?id=20120606011

With the above policy and community initiatives undertaken in Atlantic Canada and the observed demographic changes in its population, immigration will play an enhanced role in both the region's population growth and its economic development. As a result, issues relating to the social and economic impacts of immigration, which are often raised in public circles in the immigrant-abundant regions of western and central Canada, are also expected to occupy a central place in public policy discussions in this region. Some issues include the impact of immigration on the public treasury, poverty, employment and the wages of the native born, which surface more often in public debates during periods of economic downturn or stagnation. The data analysis presented in the next three chapters contributes to this debate by presenting the case for immigration and analyzing the labour market performance of immigrants over the past three decades.

3 An Example of Cooperation Between Stakeholders in the Integration of Professional Immigrants in Nova Scotia

Doctors Nova Scotia partners with Immigrant Settlement & Integration Services Nova Scotia to help ensure International Medical Graduates (IMGs), as potential future members of the association, have the resources and support they need to embark on the licensure process.

ISIS is a community organization that welcomes immigrants to Nova Scotia. It offers services and creates opportunities to help immigrants participate fully in Canadian life. Part of its focus includes supporting medical and other health care professionals.

The IMG Bridging Program at ISISNS includes the following:

- Specialized employment counselling and coaching
- Pathways to licensure—flowchart diagrams for family physicians and specialists
- English for health care professionals
- Study groups—including IMGs and Canadian-born medical students and residents
- Orientation sessions and workshops—topics include research methodologies, medical law and ethics, and tips for objective structured clinical examination (OSCE)
- Clinical skills orientation programs—innovative programs delivered in partnership with Canadian licensed physicians and the Dalhousie Faculty of Medicine Learning Resource Centre that help IMGs prepare for clinical assessments as well as effective professional practice in Canada

- Professional mentors—a partnership initiative with Doctors Nova Scotia and Dalhousie Medical School
- Observerships in family medicine practices and clinics—the latest partnership initiative involving Doctors Nova Scotia, Dalhousie and ISIS
- Learning resources—a collection of medical resources, assistance in accessing hospital and university libraries, and an electronic bookshelf through Doctors Nova Scotia and ISIS.

Source http://www.doctorsns.com/en/home/practisinginnovascotia/internationalmedicalgraduates.aspx

References

Akbari, A.H. (2009). *Socioeconomic and Demographic Profiles of Immigrants in Atlantic Canada*. Report prepared for Atlantic Canada Opportunities Agency (www.atlantic.metropolis.net).

Atlantic Institute for Market Studies (2009). *In the Media Atlantic Canada's aging population and expected labour shortages*. Retrieved October 29, 2011, from http://www.aims.ca/en/home/library/details.aspx/2746.

Atlantic Provinces Economic Council (2007a). *Atlantic Report* (Winter).

Atlantic Provinces Economic Council (2007b, April). *Where Have all Workers Gone?*

Canadian Tourism Research Institute and Conference Board of Canada (2010). *The future of Canada's tourism sector: Economic recession only a temporary reprieve from labour shortages. Canadian Tourism Human Resource Council*. Retrieved October 29, 2011, from www.cthrc.ca.

McNiven, J.D., & M. Foster. (2009, January). *The developing workforce problem: Confronting Canadian labour shortages in the coming decades*. Halifax: Atlantic Institute for Market Studies.

Newfoundland and Labrador (2007a). *All the skills to succeed: Report of the Newfoundland and Labrador skills task force report*. Retrieved October 29, 2011, from http://www.ed.gov.nl.ca/edu/publications/stf-report.pdf.

Ray Barton Associates Ltd. (2008). *Trends and patterns in skills and labour shortages*. Final report submitted to Council of Deputy Minister's Secretariat. Retrieved October 29, 2011, from http://www.comt.ca/english/LabourSkills.pdf, Ottawa.

Statistics Canada. (2011). *Estimates of total population, Canada, Provinces and Territories (Special tabulations, Demography Division)*.

Tomlin, B. (2007). *The seat shortage: Changing demographics and representation in the house of commons*. C.D. Howe Institute (e-brief, May 29, 2007).

Chapter 3
Immigration Trends in Atlantic Canada

Unlike some provinces, such as Ontario, the Atlantic provinces are immigrant scarce; immigrants comprise less than 4 % of the regional population compared to the national average of about 18 % (Table 1).

The region also receives a small share of annual immigrant inflows to Canada. In 2010, while its population was about 6.9 % of the total Canadian population, Atlantic Canada was the destination of only 2.8 % of immigrants who arrived in Canada. This was, nonetheless, the highest share since 1997. Chart 1 shows the trends in immigrant inflows into the four Atlantic Provinces for the period 1981–2010.

Nova Scotia is the most popular destination of immigrants to the region. During the mid-1990s, Nova Scotia experienced a sudden rise in its immigrant inflows, reaching a peak at about 3,600 in 1995 from about 1,500 in 1991. This dramatic rise took place soon after the first Gulf War, which resulted in the emigration of Palestinians and several expatriate groups from Kuwait and its neighbouring countries and had some specific reasons (see Box 1). Towards the end of 1990s, however, annual inflows to the province returned to their early 1990s level, averaging at about 1,700. Immigrant inflows into the other three provinces remained relatively stable during the entire 1981–2002 period.

Since 2002, Nova Scotia, New Brunswick and Prince Edward Island have been welcoming rising numbers of immigrants each year, with tiny Prince Edward Island (total resident population about 140,000) receiving the most in 2010. Growth has been slower in Newfoundland and Labrador but has picked up since 2005, increases that may be largely attributed to deliberate attempts at government and community levels to attract immigrants to, and retain them in, the region. The introduction of the Provincial Nominee Program (PNP) is one example as it allows provinces and territories to nominate immigrants who have specific qualities and skills that will contribute to the local economy. Under this agreement, each province nominates candidates who want to work and live in that province. Nominees must meet federal admissibility requirements, such as those related to health, criminality and security. As a part of the agreement, local communities can also identify an immigrant nominee to fulfill their own labour market needs.

S. A. H. Akbari, *Immigrants in Regional Labour Markets of Host Nations*,
SpringerBriefs in Population Studies, DOI: 10.1007/978-94-007-6244-2_3,

Table 1 Immigrant status of population in Atlantic Canada, 1981–2006 censuses

Census year	Total population	Non-immigrants	Immigrants	Immigrants in total population (%)
1981	2,214,140	2,130,015	84,125	3.80
1991	2,299,480	2,223,825	75,655	3.29
2001	2,258,750	2,182,790	75,960	3.36
2006	2,257,550	2,172,795	84,755	3.75

Source Statistics Canada, catalogue no. 97-557-XCB2006006. No information is available on the immigrant status of total population for the post-2006 period. Special tabulations based on the Labour Force Survey indicate that in the population aged 15 and above, immigrants comprised about 4 % in 2010

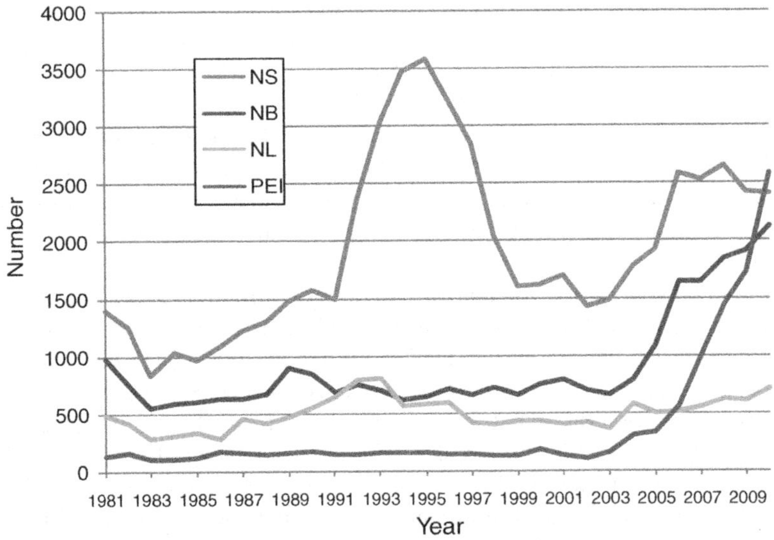

Chart 1 Immigrants (principal applicants and dependents) destined for each Atlantic province, 1981–2010. *Source* Table A5

As Table 2 shows, New Brunswick was the first province to enter such an agreement in 1999. Different provinces entered the agreement with the federal government at different dates, initially for a limited term, but now each has been granted indefinite extensions to its original agreement. The dramatic rise in immigration in Prince Edward Island (about 24 times) since 2002 (see Chart 1) was almost entirely due to the provincial nominees, who comprised 94 % of total inflows to that province in 2010, when it also received the highest number of immigrants among all provinces.

Chart 2 data show that since 2000, annual immigrant arrivals under PNP have been rising and that on average, each provincial nominee in Atlantic Canada brings about three dependents.

Table 2 Federal-provincial/territorial agreements currently in force in the Atlantic Provinces

	Date signed	Expiry date
Canada—Newfoundland and Labrador agreement	November 22, 2006, Originally signed September 1999	Indefinite
Agreement for Canada—Prince Edward Island cooperation on immigration	June 13, 2008, Originally signed March 2001	Indefinite
Agreement for Canada—Nova Scotia cooperation on immigration	September 19, 2007	Indefinite
Canada—New Brunswick Agreement on Provincial nominees	January 28, 2005; amended March 29, 2005, Originally signed February 1999	Indefinite

Source CIC. 2010 (Section 3: Federal-Provincial/Territorial Partnership)

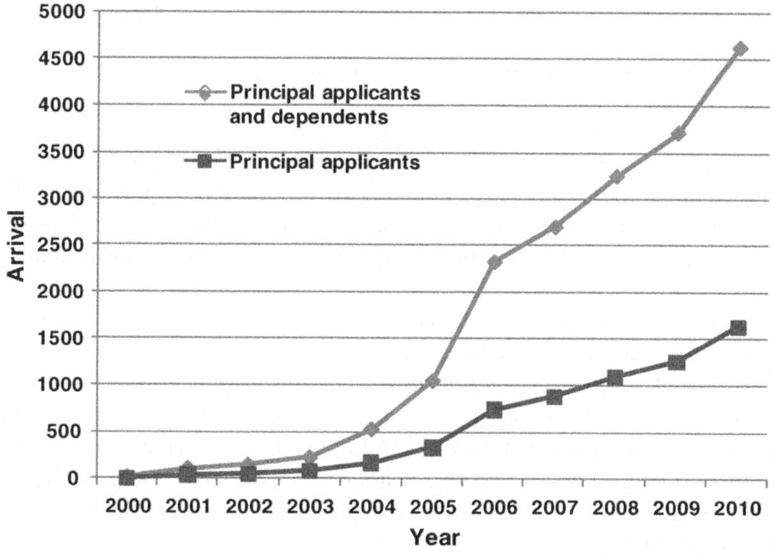

Chart 2 Annual arrivals of immigrants in Atlantic Canada under the provincial nominee program, 2000–10. *Source* Table A7

The nominee share of annual immigrant arrivals varies across provinces. Chart 3 shows the percentages of provincial nominees (principal applicants and dependents) in annual arrivals for Atlantic Canada and for each province since 2005. The region has experienced a consistent rise in its share of provincial nominees in its total immigrant arrivals since 2005. Since 2008, at least half the immigrants destined to the region came as provincial nominees. Prince Edward Island received the highest percentage of its immigrants as provincial nominees each year, totaling about 94 % in 2010, while Newfoundland and Labrador has experienced a sudden jump in provincial nominees to 30 % in the last 2 years. Most arrivals in New Brunswick since 2006 were also provincial nominees. Their percentages in Nova Scotia have been modest over the 2005–2010 period and even declined in 2010.

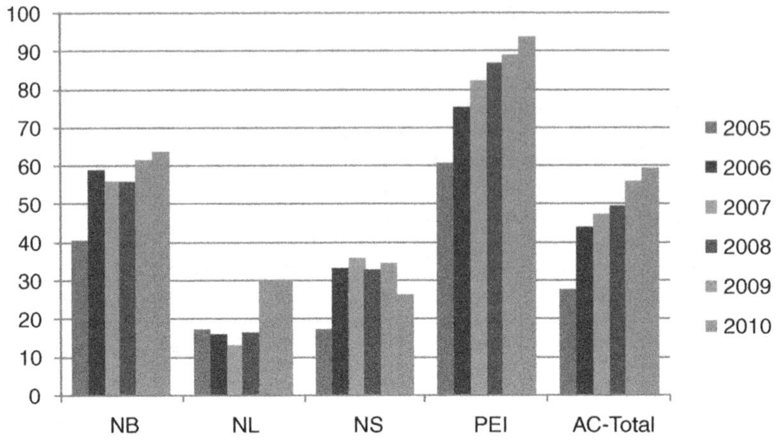

Chart 3 Annual arrivals of provincial nominees in Atlantic Canada by province, 2005–2010 (%). *Source* Table A8

In sum, the new immigration to Atlantic Canada can be viewed as more "job oriented" than in the past. Larger numbers of immigrants now arrive in Atlantic Canada with job offers in hand than in the past, mainly due to the introduction of the PNP. With the adoption of immigrant attraction and retention initiatives, the recent rising trends of immigrant arrivals are expected to be sustainable as a critical mass of immigrant population develops in the region. The current rise in immigrant retention can also be attributed to the rising inflow of provincial nominees, which a more systematic research study should further investigate by estimating an econometric model for immigrants' location choice and determining whether these rising immigrant trends would have been observed even without the introduction of the PNP. In immigrant-abundant provinces such as Ontario, the presence of an immigrant population appears to be an important source of immigrant attraction.

1 The Rise and Fall of Immigration in Nova Scotia in the 1990s

In the aftermath of the first Gulf War, many Palestinian and expatriate groups living in Kuwait and in its neighbouring countries started to leave. Rising emigration from the Middle East caught the attention of some aggressive immigration consultants, who began to promote Nova Scotia as a province that has a more conservative family lifestyle, is safer than big Canadian cities, and has nationally ranked educational institutions. As a result, immigration peaked at 3,600 in 1995 and remained close to 3,000 per year until 1997 (see Charts 1 and 2). Most immigrants from the Middle East came as business class immigrants. The federal entrepreneur program at the time had fairly relaxed requirements (start a business within 2 years and employ one Canadian). Many immigration consultants even helped clients write business plans so that they could get into the country. However, many of these immigrants were actually professionals with no prior business experience but were told by consultants that it would be easy to do business in Nova Scotia. They had the money to invest, and Nova Scotia appealed to them. However, they encountered the following problems:

- Many found that there were not as many business opportunities and settled for small retail operations, which they could not manage properly.
- Even those with some business experience had mostly done international trade (import/export business), which was not a lucrative business in Nova Scotia.
- Incomplete or erroneous information was provided by consultants about business opportunities in the province.

As a result, many immigrants started to leave the province and may have also advised potential newcomers not to immigrate to Nova Scotia. By 1999, the province had returned to its pre-1991 level of annual immigrant inflows.

In 2002, the federal government, concerned about the abuses of the system, changed the entrepreneur program drastically, requiring, among other things, recent entrepreneurial experience, a large initial investment, and more direct involvement in the business to be eligible under the program. As a result, business immigration to the province, which had already declined significantly since the 1990s, has now dropped to below its 1981 level. The federal government has also taken major steps to monitor immigration consultant activity throughout the country.

The consultants' attention was also diverted from business immigrants when Nova Scotia signed PNP agreements with the federal government in 2002.

Source Based on information collected from ISISNS.

The Rural–Urban Settlement Pattern of the Immigrant Population

Immigrants to Canada tend to prefer urban regions to rural ones. The 2001 census revealed that immigrants comprised 27 % of urban and 6 % of rural area and small town populations in Canada. By 2006, immigrants' composition of urban and rural/small town populations had risen to 30 and 7 %, respectively. The urban trend among immigrants was intensified by those immigrants who had arrived since 1986 because they made up just 1.6 % of the predominantly rural—but more than 15 % of the predominantly urban—regional populations. Between 2001 and

2006, about 31,075 new immigrants settled in rural Canada, representing only 0.5 % of all people living in rural areas and small towns in Canada. During the same period, however, about 4.3 % of those living in larger urban centres were new immigrants (Beshiri and He 2009).

Declining rural population trends can cause (1) a decline in natural resource industries, such as agriculture and mining, and (2) a decline in public and private services in rural regions, as well as (3) increased pressures on the provision of services in metropolitan (urban) areas as the rural population moves there. Federal and provincial governments have responded by adopting rural re-population strategies, one component being the initiative to attract immigrants to rural regions using PNPs and community-based initiatives as tools to achieve this goal. Some authors (for example, Broadway 2007) noted that immigrants have integrated into the economies of communities that have established local services for an immigrant population.

2 Rural Nova Scotia Faces Shortage of Lawyers

The Barristers' Society of Nova Scotia reports that rural Nova Scotia is facing a drastic lawyer shortage that will leave communities underserviced if nothing changes. The shortage is mainly related to the aging of lawyers in rural Nova Scotia. The concern has become so strong that on November 3, 2011, the Lunenburg County Barristers' Association hosted a conference in Bridgewater on the challenges and opportunities facing lawyers in rural Nova Scotia. The county will lose half of its 52 lawyers due to retirements over the next 20 years. People will have to travel to Halifax, where the service is more expensive. In addition, they will have to take time off work and spend more just getting to appointments.

Source Chronicle Herald, November 4, 2011

Chart 4 shows the urban–rural distribution of immigrant and non-immigrant populations in Atlantic Canada as revealed by the 2006 census. Following Statistics Canada definitions, a Census Metropolitan Area (CMA) and Census Agglomeration Area (CA) are considered to be urban. Based on the Metropolitan (CMA and CA) Influenced Zones (MIZ) classification system, rural regions are divided into four categories: strong MIZ, moderate MIZ, weak MIZ, and no MIZ. Each category represents a specific degree of "rurality," with strong MIZ being the least rural and no MIZ being the most rural. This categorization has proved useful for developing the profiles because they highlight differences between types of rural-based labour market integration as a basis for rurality.

To classify population into rural and urban areas, Statistics Canada uses different concepts. One of them is the Metropolitan Influenced Zone (MIZ) classification system, an approach to better differentiate areas of Canada outside of census metropolitan areas (CMA) and census agglomerations (CA). Census

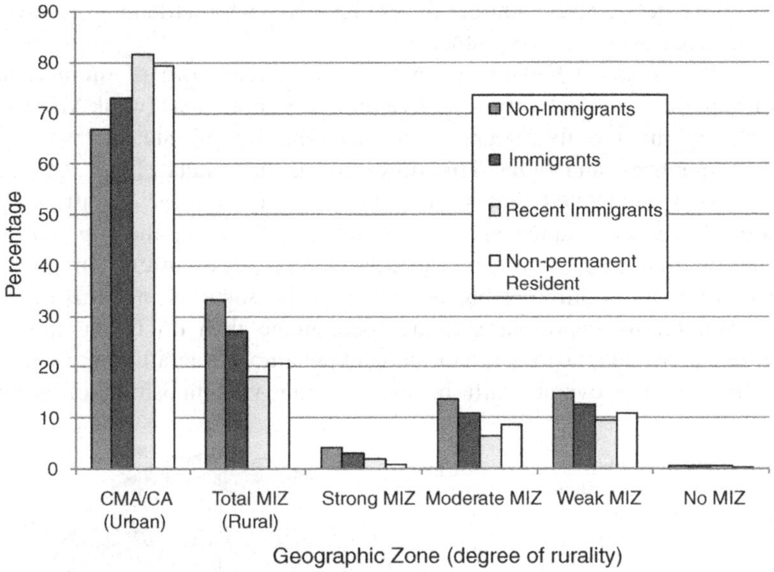

Chart 4 Population distribution in Atlantic Canada, 2006. *Source* Table A9. *Note* statistics canada has adjusted the MIZ data by redefining some cma boundaries. Therefore, these data are not comparable to data based on the 2001 census

subdivisions that lie outside these areas are classified into one of four zones of influence ranging from "strong" to "no" influence, according to the degree of influence that CMA/CAs have on them. This type of categorization has proven useful for developing the rural–urban profiles because it highlights differences between types of rural-based labour market integration as a proxy for rurality.[1]

As shown in Chart 4, about 33 % of Atlantic Canadians lived in rural areas at the time of the 2006 census (roughly the same percentage as in 2001, according to revised data). Therefore, any change in the composition of the rural population in the total population is expected to significantly impact the region's economy. As expected, a larger percentage of immigrants (73 %) than non-immigrants (67 %) lives in urban Atlantic Canada; however, about 27 % of all immigrants and 18 % of recent immigrants also choose to live in rural areas. This indicates that such areas can be destinations for new immigrants. Interestingly, about half of recent immigrants who choose to live in rural areas go to those with high degrees of rurality (i.e., weak MIZ).

The chart also shows the distribution of non-permanent residents whose likelihood to settle in rural areas is higher than that of recent immigrants. Most are temporary foreign workers recruited to work in specific jobs for which non-immigrants are not available. These jobs could be in fishing, agriculture or care-

[1] Chuck et al. (2000) give a detailed explanation of the MIZ classification system.

giving to the elderly. These non-permanent residents who settle in urban areas also include international university students.

For more insights, Charts 5 and 6 provide a breakdown of immigrants and recent immigrants, respectively, who live in highly rural areas (weak MIZ) and are organized by the top five source countries: the United States, three western European countries, and India. This differs from the results of the 2001 census, which showed the United States and four western European countries (United Kingdom, Germany, France and the Netherlands) among the top five source countries of immigrants living in weak MIZ areas. The current composition of recent immigrants is different but nevertheless the same as in the 2001 census. Chinese immigrants represent a higher percentage than do Indian immigrants among recent arrivals; however, their percentage among overall immigrants is still lower. In sum, the overall shift in source country composition of immigrant

Chart 5 Immigrants in weak MIZ of Atlantic Canada by source country, 2006

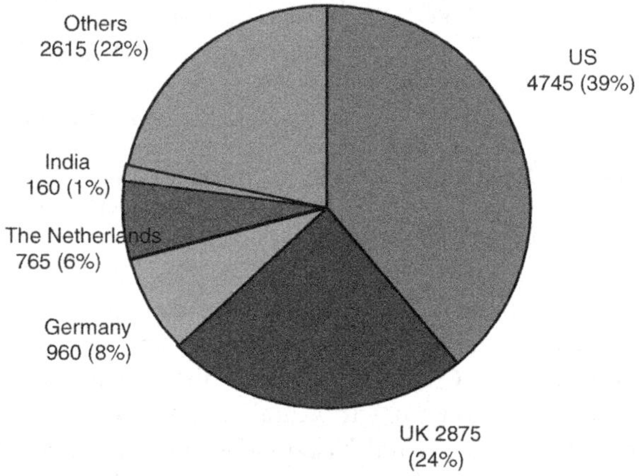

Chart 6 Recent immigrants in weak MIZ of Atlantic Canada by source country, 2006. *Source* Table A10. *Note* The top five source countries are considered

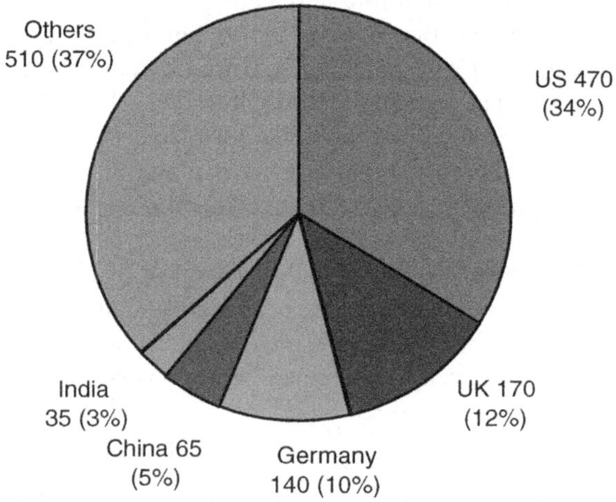

inflows to Canada (to be discussed in later in the chapter) is also felt in rural regions but perhaps not so remarkably as in urban regions, as evidenced by the continued dominance of Americans and Europeans among recent arrivals heading to rural areas.

No breakdown of rural and urban populations by immigrant status for the post-2006 period is available, but the Labour Force Survey data do allow a breakdown of labour force by urban (CMA/CA) and rural (non-CMA/CA) residence criteria based on immigrant status. Chapter 4, which discusses the characteristics of the immigrant labour force living in Atlantic Canada, also discusses this breakdown.

Age Distribution Among New Immigrants

It is usually the young who migrate because young individuals can reap the benefits of their migration investment for a longer time at their destination than can the old. The young usually are also healthier and more energetic.

Chart 7 confirms that international immigrants to Atlantic Canada are indeed younger than non-immigrants at the time of arrival (that is, within a given five-year period). Most immigrants and non-immigrants are aged 25–44, the prime working age group. However, the number of non-immigrants in this age group as a share of total population has been declining since 1991, and by 2006, the same share fell below that of the 45–64 age group, which is considered to be close to retirement

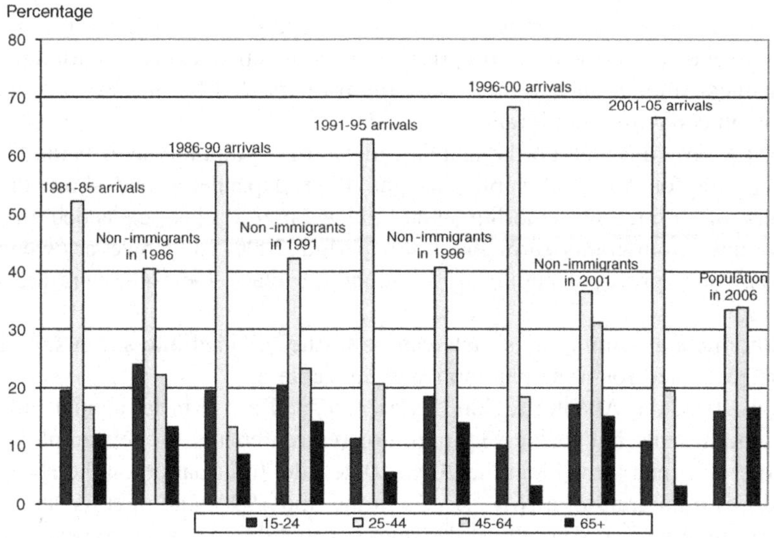

Chart 7 Age distribution of new immigrants (principal applicants) and non-immigrants in atlantic canada, 1981–2006. *Source* Table A11. *Note* Data reported for population aged 15 and over

age. On the other hand, since 1991, more than 60 % of immigrants arriving in Atlantic Canada have been in their prime working age.

Chart 7 also reveals the overall aging trend of the Atlantic population. The share of non-immigrants aged 65 and over rose throughout the 1981–2006 period, while the share of youth (aged 15–24) declined. These trends partly reflect declining birth rates, as discussed in Chap. 2. In addition, more and more immigrants are arriving in this region with fewer elderly. The share of youth arrivals listed is also lower because these data list only principal applicants.

The results of this section indicate that immigration can be used to (1) meet the labour shortages in the region, and (2) reverse the aging trend in the Atlantic population. However, present levels of immigration are too low to help offset the aging trend of the region's population. Therefore, research is needed to determine what level of immigration would reverse this trend among Atlantic Canadians over the next 10 years.

Composition of Immigrant Classes

Immigrants come to Canada under different classes, and applications filed under each class are evaluated according to different eligibility rules. ***Refugee class*** immigration reflects Canada's commitment to humanitarian principles, while ***family class*** immigration is intended to foster family re-unification, that is, facilitating the process of individuals entering Canada to join close relatives who are already permanent residents/citizens. ***Economic immigrants***, on the other hand, are those expected to make a direct, positive economic contribution to Canada through the skills, expertise, entrepreneurship, and/or capital they bring with them. The economic class category itself comprises two major sub-categories: skilled workers and business immigrants. In this study, the primary (but by no means exclusive) focus is on economic immigrants.

Chart 8 provides data on the annual inflows of immigrants in Atlantic Canada by class are for principal applicants and their dependents and show that the numbers of immigrants entering under the economic class exceeded those of family class immigrants throughout the period 2000–10. However, when one considers only principal applicants in economic class, their numbers are much smaller.

Economic class immigrants' data can be further divided into seven sub-classes and are presented for principal applicants in Table 3.

Since 2003, the Atlantic region has experienced a rise in its annual inflow of skilled immigrants, but business immigration (entrepreneurs, the self-employed and investors) declined throughout the 2000–10 decade. Information on the abrupt rise in business immigration in Nova Scotia in the mid-1990s and then its sudden fall was presented earlier. Therefore, the region's inability to attract business immigrants directly from abroad should be investigated and addressed by policy makers. As shown by Akbari (2009), business immigrants also moved to Atlantic Canada from other regions of Canada. For example, during 1990–2004, the region attracted

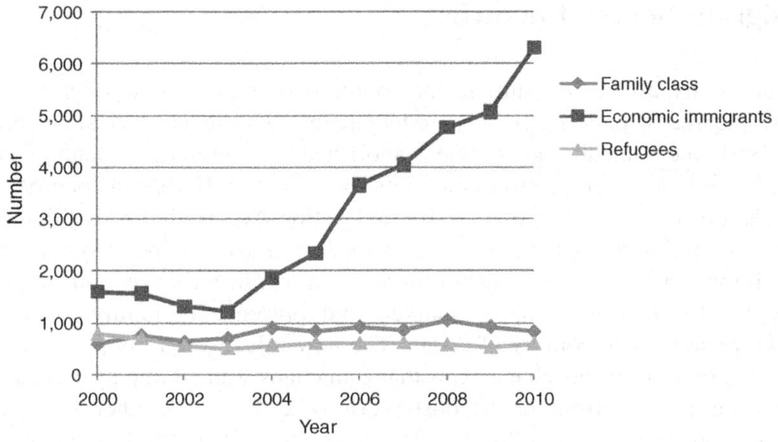

Chart 8 Annual arrival of immigrants in atlantic canada by class, 2000–10. *Source* Table A6

Table 3 Economic immigrants (principal applicants) destined to Atlantic Canada by type, 2000–2010

	2000	2001	2002	2003	2004	2005	2006	2007	2008	2009	2010
Entrepreneurs	90	50	20	15	30	15	10	5	10	***	***
Self-employed	20	25	20	15	10	***	15	***	***	***	***
Investors	25	20	10	15	20	15	15	25	10	20	40
Skilled workers	470	500	405	320	445	465	520	525	600	475	515
Canadian experience class	0	0	0	0	0	0	0	0	0	25	45
Provincial/territorial nominees	10	50	60	90	180	345	750	890	1,105	1,265	1,640
Live-in caregivers	5	***	***	10	15	10	10	20	20	30	45
TOTAL	615	635	525	460	700	850	1,325	1,475	1,740	1,820	2,290

Source Based on special tabulations by CIC, Facts and Figures, 2010
"***" means data not available due to suppression

115 business immigrants from other regions of Canada who had landed there during that period, a number equal to about 63 % of those directly destined to the region. However, as is observed in Table 3, during the 2003–07 period, the region attracted only 20 business immigrants from other regions and lost 40 to them.

The fall in the number of business immigrants to the region can be partly attributed to the termination of the business mentorship program in Nova Scotia. The province ended up settling a $30 million lawsuit with participants in the program, who had paid $130,000 each to come to Nova Scotia following the promise of six months' employment. Many ended up being placed outside of their field of expertise in a job lasting only as long as the government subsidy—which the immigrant had actually paid for—to the employer (Beswick 2011).

Immigrant Source Countries

Canada has experienced a shift in the source countries of its immigrants in the post-World War II period, as discussed by Green (1976) in a historical perspective.

In 1961, the Canadian government abolished the "preferred country" clause, which had given preference to immigrants from western European countries. This clause had formed the basis of a 1910 Immigration Act. With the abolition of this clause, all immigrant applications are now evaluated using a "point system" under which importance is given to such criteria as an applicant's age, education, suitability for the Canadian labour market, and presence of family members in Canada, regardless of country of origin. The new rules were fully promulgated in 1967. One reason for this change was that Canadians wanted to play a greater role on the international front in the post-World War II era. Another was that the economic prosperity that followed soon after the war resulted in an increased demand for skilled labour.

The period of the early 1960s was also a time when economic conditions in Europe, adversely affected by World War II, had begun to improve. More labour was in demand and per capita incomes were rising. As a result, immigration from Europe to North America generally slowed down. Immigration slowed further with the formation of the European Union and the re-unification of Germany, which allowed for great mobility of labour within Europe.

A consequence of the above-noted changes, as well as of the greater mobility of labour in a globalized world and continuing political discourse in the third world countries, has been a shift in the source countries of Canada's immigrant inflows from the countries of western Europe to those of Asia, Africa, and South and Central America over the past three decades.

In Atlantic Canada, change in the source countries of immigrants became more prominent only in the 1990s, when China and some Middle Eastern countries entered the list of top five immigrant source countries (Table 4). The entry of Middle Eastern countries in the 1990s can be attributed to the first Gulf War, which adversely affected some groups and prompted them to leave their countries of residence, and to some private immigrant recruitment activities undertaken in Nova Scotia (see Box 1 above). These countries also became more prominent in the list of top five source countries in the 1990s, when both principal applicants and dependents were considered, likely due to the larger family sizes of immigrants originating there.

In recent years, the prominence of Middle Eastern immigrants among the top five source countries has diminished; instead, China has consistently been on the top of the list. The United States and the United Kingdom (despite the general fall in immigrant inflows from Western Europe to Canada since the 1970s) are permanent members of this list, probably due to their traditional ties to Canadian society, shared history, common language and close proximity. The presence of the large immigrant population that came from those countries in the past may be

Table 4 Top five source countries of immigrants destined for Atlantic Canada by year, 1981–2010

Period	1981–1985		1986–1990		1991–1995		1996–2001		2002–2006		2007–08		2009		2010	
	Country	Count	Country	Count	Country	Count	Country	Count	Country	Count	Country	Count	Country	Count	Country	Count
Principal applicants and dependent																
Rank																
1	USA	3,284	USA	2,620	Egypt	1,519	China	1,948	China	1,919	China	2,266	China	1,645	China	2,680
2	UK	1,931	UK	1,606	USA	1,510	Kuwait	1,629	USA	1,488	Korea	1,347	UK	625	UK	695
3	Vietnam	748	Poland	693	Kuwait	1,247	Jordan	828	UK	1,184	UK	1,088	Korea	530	Korea	370
4	Poland	331	Vietnam	617	Hong Kong	1,224	Korea	783	Korea	1,091	US	928	USA	425	Iran	365
5	Germany	321	Lebanon	444	Saudi Arabia	887	Saudi Arabia	677	Iran	246	Iran	543	Iran	340	USA	315
Total for 5 countries		6,615		5,980		6,387		5,865		5,928		6,172		3,565		4,425
Total arrivals		11,398		13,340		21,495		20,841		17,886		12,297		6,660		7,830

Source From 1981–2004, Permanent Resident Data System (PRDS) micro-data as provided to AMC under contract with CIC. Principal applicant is based on variable "f_stat2", and source country is based on variable "f_clpr". The 2004–10 data are based on special tabulations obtained from CIC another factor attracting them to the region. The inclusion of South Korea and Iran on the list of top five source countries of immigrants is also worth noting.

Finally, it is also noteworthy that the top five source countries of immigrants have formed less than half of total immigrant inflows to the region for most of the period under study. This means that immigrants to the region also arrive in small numbers from diverse source countries. For example, in the last 2 years, the top five source countries comprised more than 50 % of all immigration, mainly due to a sudden jump in the inflow of Chinese immigrants; however, in 2010, the Philippines was the largest immigrant source country in Canada, followed by China.

Immigrant Retention in Atlantic Canada

It is well known that most immigrants who arrive in Canada aim to settle in Alberta, British Columbia, Ontario and Quebec because of, among other reasons, greater economic opportunities, larger immigrant populations from the source countries of recent immigrants, and larger family networks than those offered by smaller provinces. Even many immigrants whose initial destination is a different province eventually end up in one of these four major provinces. As a result, smaller provinces, such as those of Atlantic Canada that wish to increase their share of total immigrant inflows, face the two-fold challenge of not only attracting, but also retaining, new immigrants.

In this study, immigrant retention rates are calculated using two sources: (1) census data combined with annual arrivals data provided by CIC, and (2) the latest available Longitudinal Immigrant Data Base (for the year 2008) produced by Statistics Canada and CIC. These data, which also help calculate retention rates for each class of immigrants, will be described shortly.

Chart 9 shows the 5-year immigrant retention rates in the Atlantic region. Calculations are based, from census data, on the number of recent immigrants living in the region as a percentage of those destined to the region in a given five-year period (some of these latter data were purchased from CIC and some are based on Akbari 2009). The number listed in Chart 9 as living in the region includes those initially destined for both Atlantic Canada and for provinces outside the region.

Atlantic Canada had difficulty retaining its new immigrants from the mid-1980s to the mid-1990s, when the retention rate among recent immigrants plunged sharply from more than 70 % to about 45 %, which can likely be attributed to the changing composition of immigrant source countries. More immigrants now arrive in Atlantic Canada from non-Western European countries and with different ethnicity and language. Once they arrive, they appear to move toward the four major immigrant-receiving provinces. This onward movement could be for some or all of the following reasons: (1) larger numbers of immigrants originating from their own source countries live in those provinces; (2) larger provinces offer greater economic opportunities; and (3) immigrants arriving in Atlantic Canada face such

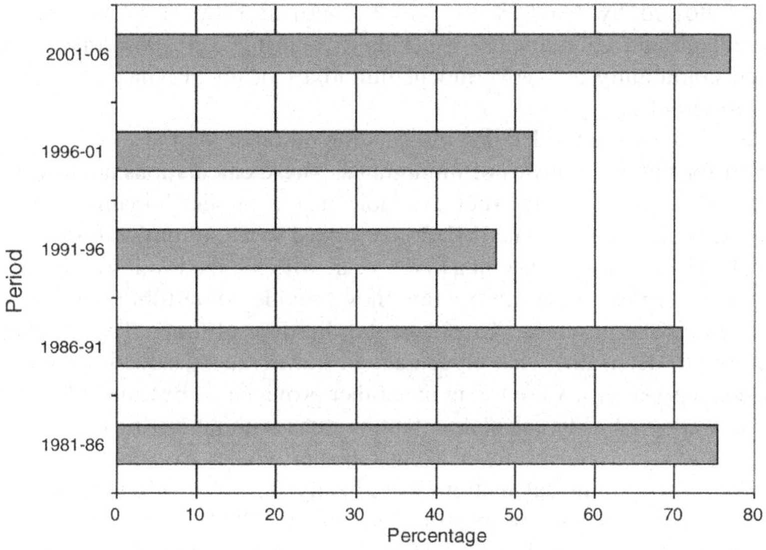

Chart 9 Five-year immigrant retention rates in Atlantic Canada, 19812006. *Source* Table A12

labour market barriers as language difficulty or discrimination. It would be useful to investigate these reasons in a systematic research study.

Therefore, in recent years, Atlantic Canadian provinces have begun to adopt new initiatives to increase their immigrant retention rates. For example, Nova Scotia's immigration strategy has a stated objective of raising its retention rate to 70 % by 2020. Newfoundland and Labrador has also stated this objective in its recently announced immigration strategy, as have New Brunswick and Prince Edward Island. To pursue this objective, all provinces now actively provide, among other things, job counseling, help in settlement, and English language training to new immigrants, some provided through immigrant settlement agencies, which receive funding from federal and provincial governments. The need to create more welcoming communities to attract more immigrants to the region is also emphasized in policy circles. Each province has also renewed its provincial nominee agreement with the federal government, as discussed earlier in this report.[2]

These initiatives have started to show results, as was reflected in the immigrant retention rates for 2001–2006 (Chart 9). Atlantic Canada's retention rate, which had started to rise in the late 1990s, showed a sharp rise at the turn of the present century, and about 77 % of immigrants who arrived in the region during 2001–06 were still living there at the time of the 2006 census. This retention rate exceeds even that of the early 1980s. In fact, all four provinces have increased their retention over the 2001–06 period, with New Brunswick retaining the most

[2] Each province's immigration strategy can be found on its government's web site.

(91 %), followed by Nova Scotia (75 %), Prince Edward Island (72 %) and Newfoundland and Labrador (59 %). This rise during 2001–06 primarily reflects increased community and government initiatives in all provinces to attract and retain immigrants.

Table 5 provides more detailed information on retention rates which have been calculated for different classes of immigrants. These calculations are based on the 2008 IMDB summary tables (not available in the public domain) compiled by Statistics Canada and CIC. The IMDB is designed to track immigrants destined for a particular Canadian province in a given year, who are then followed for 15 years in Canada according to the information they provide while filing their tax return, which includes the province of residence at the time of filing. This database can also be used to determine what percentage of immigrants destined for a particular province in a given year were living in another province at the time of filing. While census data allowed us to calculate retention rates without distinguishing between immigrants who may have moved from other provinces to Atlantic Canada, the IMDB data also permit calculations based only on those originally destined to Atlantic Canada and who were still there in the 2008 taxation year. Separate calculations are also possible for refugees, economic immigrants and family class immigrants, among others. Economic immigrants are further divided into business and skilled immigrants (principal applicants only). Only region-wide data are available due to the small numbers in each Atlantic province but are further divided into different categories of immigrants who arrived in Canada over the period 1993–2007.

Table 5 data indicate that most immigrants destined for Atlantic Canada (including those originally destined to the region, and those who came from other parts of Canada) during 1993–2007 and who were still living there in 2008 had come as economic immigrants, while those who had come as refugees or family class immigrants were evenly divided. Overall, about 57 % of immigrants originally destined to Atlantic Canada were living outside the region in the 2008 taxation year. However, about 17 % of this loss was offset by in-migration of those originally destined to other Canadian provinces, thereby reducing the net loss of immigrants in Atlantic Canada to about 40 %. Economic immigrants (business, skilled and other economic) were the most likely to move out, while family class immigrants were the least likely. However, economic immigrants also move in from other regions of Canada, thereby reducing Atlantic Canada's net out-migration rate.

To summarize, in 2008, Atlantic Canada's net immigrant retention rate was about 60 % of those originally destined to the region over the 15-year period ending in 2007.

Among those who arrived in Canada during 2003–07, the net retention rate rose to 80 % in 2008 (evident by "net-outmigration" rate results). During this period, the out-migration rate among all classes was below 50 % and was lower than for the entire 15-year period (reported in the top panel of Table 5). The in-migration rate (inflows from outside of Atlantic Canada) also fell among this entry cohort but more slowly than the fall in out-migration rate. The in-migration of skilled immigrants remained constant after a significant rise during 1999–2004, as shown

Table 5 Migration of immigrants destined for the Atlantic provinces

	Number of immigrants					As percentage of numbers destined		
	Resident in 2008	Destined at landing	Out-migration[a]	In-migration[b]	Net out-migration	Out-migration	In-migration	Net out-migration
Period of Arrival in Canada: 1993–2007								
By category								
Family	4,825	5,545	1,775	1055	720	32.01	19.03	12.98
Business (P.A.)	620	1,695	1,185	110	1,075	69.91	6.49	63.42
Skilled (P.A.)	3,480	4,400	2,570	1650	920	58.41	37.50	20.91
Other econ	4,265	8,790	6,175	1650	4,525	70.25	18.77	51.48
Refugee	4,115	8,735	5,045	425	4,620	57.76	4.87	52.89
Other	1,055	1,510	735	280	455	48.68	18.54	30.13
Overall	18,365	30,685	17,490	5,170	12,320	57.00	16.85	40.15
Period of Arrival in Canada: 2003–2007								
By category								
Family	2,155	2,350	520	325	195	22.13	13.83	8.30
Business (P.A.)	115	135	40	20	20	29.63	14.81	14.81
Skilled (P.A.)	1,495	1,505	575	565	10	38.21	37.54	0.66
Other econ	1,395	1,385	465	475	−10	33.57	34.30	−0.72
Refugee	3,090	5,125	2285	250	2,035	44.59	4.88	39.71
Other	550	600	145	95	50	24.17	15.83	8.33
Overall	8,800	11,100	4,030	1,730	2,300	36.31	15.59	20.72

[a] Those originally destined to Atlantic Canada now residing in other provinces
[b] Those originally destined to a province outside of Atlantic Canada now residing in an Atlantic province. *P.A.* Principal applicant
Source Longitudinal Immigration Database (IMDB) Summary Tables, 2008

by Akbari (2009). Anecdotal evidence suggests that the Defined License Program (DLP) for International Medical Graduates (IMGs) introduced by Nova Scotia attracted immigrant medical professionals to that province from the province of Ontario (see Box 3). The DLP creates an incentive for an IMG to stay in the province until he/she obtains the license. Basky et al. (2007) have shown that about half of IMGs (as well as Canadian medical graduates) leave Newfoundland and Labrador within 7 months of obtaining their license. However, in 2010, about 40 % of the licensed physicians in that province had obtained their medical degree from a university outside Canada. Thus, IMGs do, in the short run, help meet the need for doctors. In fact, 267 licenses were issued to IMGs in that province in 2010 (College of Physicians and Surgeons, NL 2010).

In sum, both census- and IMDB-based data indicate a rise in immigrant net retention in Atlantic Canada (77 % among all immigrants and up to 80 % among tax filers) since the turn of this century. Notably, however, if one considers only those immigrant tax filers originally destined to Atlantic Canada, the retention rates are

3 Attraction and Retention of International Medical Graduates in Atlantic Canada

Meeting the shortage of doctors in Nova Scotia

To meet its shortage of physicians, Nova Scotia uses the Defined License Program. A defined license is issued to international medical graduates (IMGs) to practice as physicians with some conditions. If a doctor has a recognized international medical degree and some postgraduate training and certification, the Licensing Medical Council of Canada (LMCC) mandates the qualifications for physicians after they graduate from medical school. They also need a certification in family medicine or in a specialty. All applicants in Nova Scotia are assessed by a credential assessment committee set up by the College of Physicians and Surgeons of Nova Scotia (CPSNS).

Some conditions in a "defined license" include the need to have a sponsor and a mentor. A sponsor can be the chief of staff at the district health authority, while the mentor is usually a local doctor with the same specialty. IMGs are also given a specified period of time to get Canadian qualifications.

The CPSNS placed 42 physicians in rural Nova Scotia between 2005 and 2010.

As a general rule, internationally trained doctors from British Commonwealth countries have an advantage over physicians trained in South Asia, the Middle East and Eastern Europe.

Source Colchester Regional Development Agency (2009) and College of Physicians and Surgeons of Nova Scotia (2010).

Recruitment in Prince Edward Island

The Internationally Educated Health Professionals (IEHP) program aims to enhance the ability to recruit physicians to PEI through assistance in the assessment of credentials of IMGs. PEI partners with Nova Scotia in assessing IMGs and identifying the required deficiencies in credentials. This partnership also includes the development of a physician-led mentorship program for physicians. The retention initiatives in the province include regular follow-up with new physicians to ensure a positive experience integrating into the medical and social community, a coordinated approach for physicians to bring forward their concerns related to their adjustment in their professional and social life, and recognition for levels of long service and acquisition of special skills.

Source Prince Edward Island (2006).

Retention of IMGs in Newfoundland and Labrador

One challenge to the recruitment and retention of new physicians in the province of Newfoundland and Labrador has been that of assessing the ability and skills of IMGs who do not initially meet the licensing requirement but who may become eligible for licensure after a period of assessment and training. For general practitioners, the College of Physicians and Surgeons (CPSNL) relies on the Clinical Assessment and Training program of the Faculty of Medicine, Memorial University, to provide assessment and training. The program is delivered at Western Memorial Regional Hospital.

Source College of Physicians and Surgeons, Newfoundland and Labrador (2010).

Attraction and Retention in New Brunswick

The province of New Brunswick has had a general physician recruitment strategy in place since 1999. It focuses on recruiting family practitioners and specialists to the province; however, only a restricted license is available, and in very limited circumstances, for certain foreign-trained physicians for direct employment by a hospital or an institution.

Source Government of New Brunswick (www.gnb.ca)

lower (only 43 % for the 15-year period and 64 % for the 5-year period, both ending in 2008).

The IMDB-based results are different from those that are census based because (1) census data do not distinguish between principal applicants and their dependents, while IMDB data consider only principal applicants within the economic category, and (2) as noted above, IMDB data are only for those who filed tax returns in 2008. In short, the IMDB data represent a smaller sample of immigrants than do census data.

Education Levels Among Recent Immigrants

Education is an important predictor of economic success. Research has shown that individuals with higher education levels generally experience lower unemployment rates and earn higher incomes than those with lower education levels.

With the introduction of the "point system" in 1967, education became a key requirement for the success of an immigrant's application filed under economic class. As a result, educational levels among annual immigrant inflows to Canada have risen. Based on Charts 10 and 11, education levels among immigrant arrivals in Atlantic Canada clearly have also shown a rising trend.

Chart 10 shows declining percentages until 2006 among immigrants who arrived with only a high school or lower level of education. The same trend is observed among non-immigrants. However, throughout the period, lower percentages of new arrivals had only high school or lower education than did non-immigrants.

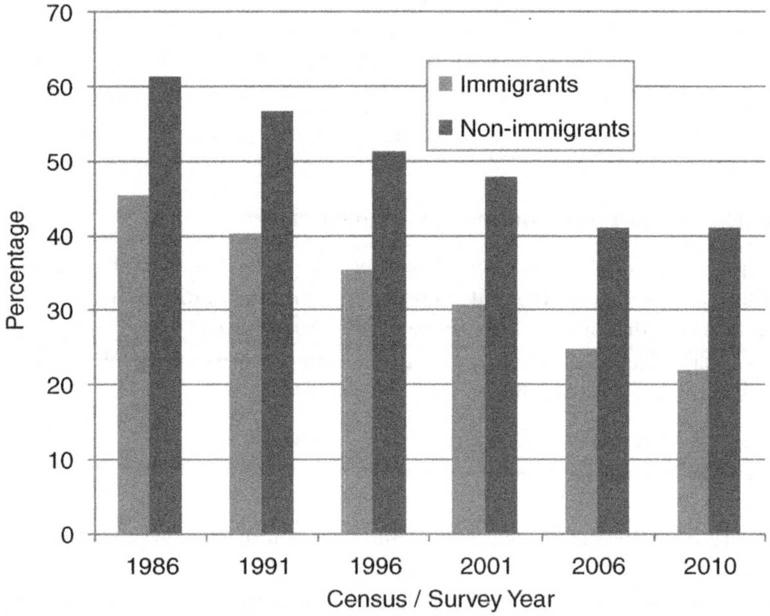

Chart 10 Immigrants destined for Atlantic Canada and non-immigrants with high school or less education, 1986–2006. *Source* and *note* Table A13. These data are for those who arrived within 5 years of the census/survey year and aged 25 and older at the time of the census. The 2010 non-immigrant data are based on the Labour Force Survey, 2010

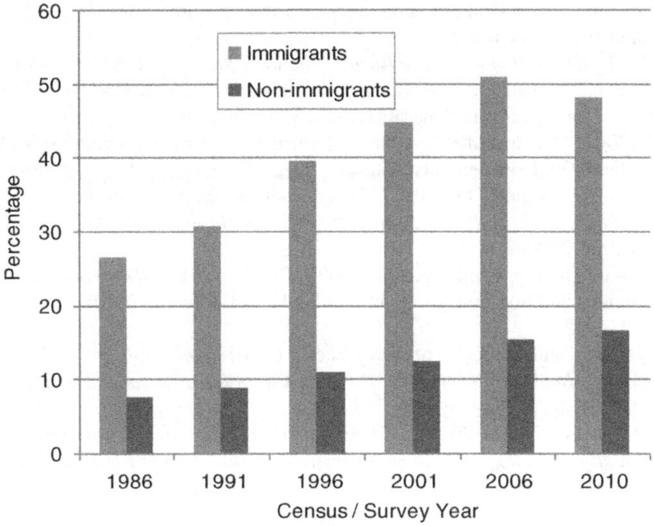

Chart 11 Immigrants destined for Atlantic Canada and non-immigrants with a university degree, 1986–2010. *Source* and *note* See note for Chart 10

On the other hand, Chart 11 shows a sharp increase in the percentage of university degree holders among new immigrant arrivals. Although a similar trend is observed for non-immigrants, their percentages rose slowly and were below those of recent arrivals throughout the period. During the last 5-year period (2006–2010), a slightly lower percentage of immigrants arrived with a university degree than did in the previous 5-year period (48 versus 51 %), possibly because of the rising number of provincial nominees in annual inflows who may have been selected as potential business investors.

Higher education levels among new immigrant arrivals can predict their labour market success only if their educational credentials are recognized in Canadian labour markets. Studies conducted by Statistics Canada have shown that an average new immigrant arriving in Canada faces difficulty finding employment because his/her foreign educational credentials are not recognized in labour markets. This adversely affects his/her labour market performance, which makes it necessary here to analyze data on the labour market performance of recent immigrants. This analysis will be undertaken in Chap. 4 using descriptive statistical tools.

References

Akbari, A. H. (2009). *Socioeconomic and Demographic Profiles of Immigrants in Atlantic Canada Report prepared for Atlantic Canada Opportunities Agency.* (www.atlantic.metropolis.net).

Basky, G., Mathews, M., Edwards, A., & Rourke, J. (2007). Do They Stay or Do They Go? Retention of Provisionally Licensed International Medical Graduates in Newfoundland and

Labrador. St. John's: Division of Community Health and Humanities, Memorial University of Newfoundland [summary report].
Beshiri, R., & He J. (2009). *Immigrants in Rural Canada: 2006* (Statistics Canada Cat # 21-006-X).
Beswick, A. (2011). *"Wanted: New Nova Scotians." Chronicle Herald* October 17. Retrieved Oct 31, 2011 from http://thechronicleherald.ca/news/25708-wanted-new-nova-scotians.
Broadway, M. (2007). Meatpacking and the transformation of rural communities: A comparison of brooks, Alberta and Garden City, Kansas. *Rural Sociology, 72*(4), 560–582.
Chuck, M., Puderer H., & Janes D. (2000). *Census metropolitan area and census agglomeration influenced zones (MIZ): A description of the methodology, statistics canada* Cat # 92F0138MIE, no. 2000–2.
Colchester Regional Development Agency (2009). *Colchester-CORDA Times*. (1st Quarter).
College of Physicians and Surgeons of Newfoundland and Labrador (2010). *Annual Report* (Saint John's).
College of Physicians and Surgeons of Nova Scotia (2010). *Annual Report* (Halifax).
Green, A. G. (1976). *Immigration and the Postwar Canadian Economy*. Toronto: Macmillan of Canada.
Prince Edward Island (2006). *Prince Edward Island Physician Recruitment/Retention and Medical Education Strategy, 2006 Government of Prince Edward Island* (Charlottetown).

Chapter 4
Immigrants in the Labour Force of Atlantic Canada

Declining population growth in Atlantic Canada has increased the role of immigration in the growth of its labour force. As Chart 1 shows, participation of new immigrants caused the Atlantic labour force to grow by only about 5 % in the early 1980s. During 1996–2001, however, new immigrants increased the region's labour force by about 45 %, some nine times their earlier contribution. While this contribution was lower during 2006–2010, it still reached 34 %. However, the composition of immigrants in the total labour force declined between 1981 and 2010, from 4.5 to 3.7 %, while increasing nationally from 19 to 21 % (data not shown here).

Table 1 provides the breakdown of Atlantic Canada's labour force by immigrant status, gender and unemployment status in 2010. In addition to providing data for the regional labour force, Table 1 also gives a breakdown by rural and urban regions.

Overall, only 3.7 % of labour force in Atlantic Canada comprises immigrants, while nationally, this percentage is about 20 %. Only about one-quarter of immigrant workers are in rural regions, where they comprise of about 2.3 % of the labour force. Around 40 % of non-immigrant workers are in the rural labour force. More men work in the rural and urban labour force, but the percentage distribution does not vary significantly between them. Also, both rural and urban unemployment rates are lower among immigrants than among non-immigrants, probably indicating immigrants' location choice in Atlantic Canada based on job availability. Recent immigrants have a higher unemployment rate, mainly due to their shorter length of stay (less than 1 year for some).

Labour Market Performance of Immigrants

Three important indicators of the labour market performance of any demographic group include their labour force participation rates, unemployment rates, and earnings, which are analyzed below for all immigrants and recent immigrants and compared with those for non-immigrants.

S. A. H. Akbari, *Immigrants in Regional Labour Markets of Host Nations*, SpringerBriefs in Population Studies, DOI: 10.1007/978-94-007-6244-2_4,

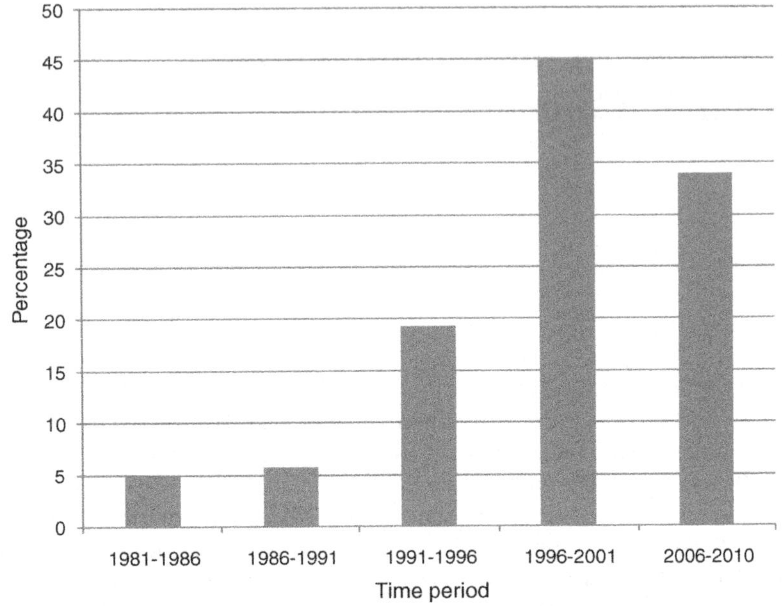

Chart 1 Immigrants' contribution to Atlantic Canada's labour force growth rate, 1981–2010. *Source and note* Table A14. Data for 2001–2006 could not be obtained. The labour force in 2010 was 1.2 million in Atlantic Canada, of which 45,400 were immigrants

Table 1 Labour force and unemployment rates, immigrants and non-immigrants by rural and urban regions, Atlantic Canada, 2010

	Rural (Non-CMA/CA)			Urban (CMA/CA)			Regional		
	Male	Female	Total	Male	Female	Total	Male	Female	Total
Labor force (Thousands)									
Immigrants	5.9	4.9	10.8	19.4	15.2	34.6	25.2	20.1	45.4
Recent immigrants	0.9	0.5	1.4	4.7	3.0	7.7	5.5	3.6	9.1
Non-immigrants	236.7	215.8	452.5	361.5	354.7	716.2	598.2	570.5	1168.7
Unemployment rate (%)									
Immigrants	10.2	–	7.4	8.2	7.2	7.8	8.7	6.5	7.7
Recent immigrants	–	–	–	14.9	16.7	15.6	14.5	13.9	14.3
Non-immigrants	16.6	11.8	14.3	9.7	6.9	8.3	12.4	8.7	10.6

Source Statistics Canada, Labour Force Survey, 2010, special tabulations

Labour Force Participation Rates

As was noted in Chap. 3, immigrants are generally younger than the resident population at the time of their arrival in Atlantic Canada. Also, most new immigrants of working age either have a job offer before arrival or become a member of

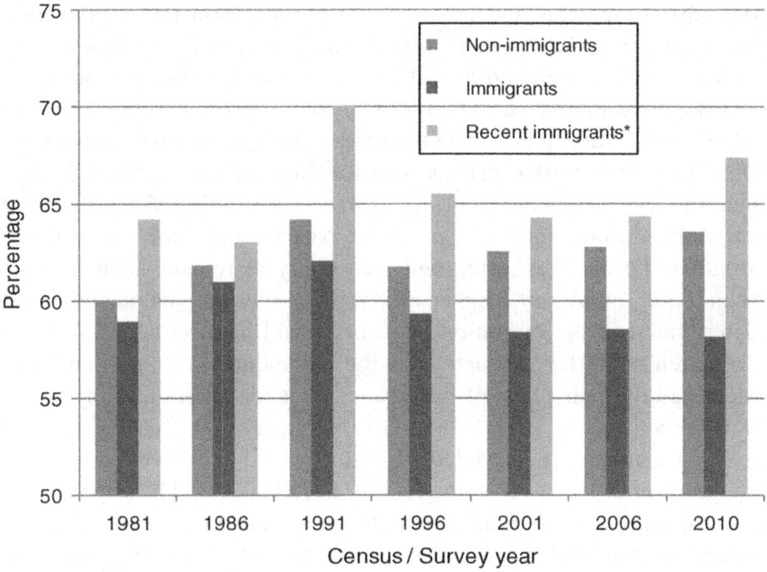

Chart 2 Labour force participation rates among immigrants and non-immigrants in Atlantic Canada, 1981–2010. *Source and note* Table A15. *Those who arrived within the past 5 years of the census/survey date. Labour force activity data based on a census are reported for the year before the census year

the labour force by looking for a job immediately after arrival, thereby increasing the region's labour force. Chart 2 confirms a higher labour force participation rate among recent immigrants (those who arrived within 5 years of the corresponding census and survey year) than among resident non-immigrants throughout the period 1981–2010. However, participation rates among recent immigrants declined substantially between 1991 and 1996 (70–66 %) and then again between 1996 and 2001 (down to 64 %) but have stayed constant since then. One reason for this decline may be that a large number of immigrants arrived toward the end of that 1991–1995 period (mostly in 1995) and thereafter. Therefore, by the 1996 census, these newcomers were still adjusting to the labour market in Atlantic Canada. Another reason for the decline in participation rates after 1996 may be that most such immigrants arrived from the Middle East and tended to have large families with many youth who, instead of entering the labour force, enrolled in secondary and post-secondary educational institutions. There is also some evidence, based on observations in Halifax, that (1) the enrollment of immigrant students from the Middle East increased in Nova Scotia universities during 1996–2001, and (2) in many Middle Eastern families, the family head may have returned to the country of origin for employment. A more formal research study should investigate these possibilities. Participation rates among all population groups remained unchanged at 1996–2001 levels during 2001–2006. After that, the rate rose among recent immigrants and non-immigrants, possibly due to a greater proportion of provincial nominees and skill class immigrants in annual arrivals.

Chart 2 also shows that the labour force participation rate among the entire immigrant population was lower than that among non-immigrants throughout the period, which can be attributed to differences in the age distributions of the two groups. Most immigrants in Atlantic Canada arrived before 2001, and the composition of the retirement-aged population (those over 65) is higher among them. One study, using 2006 census data for Nova Scotia (Akbari 2010), found that when the age distribution of immigrants was adjusted to match that of non-immigrants, immigrants' labour force participation rate exceeded that of non-immigrants because an immigrant was likely to remain in the labour force longer than a non-immigrant. Higher motivation to join the labour force, which is also a determinant of the migration decision, could be a rationale for this result. Further research could be conducted for the other three provinces and for immigrants identified by their place of birth to account for differences in social norms and cultural practices.

Table 2 presents the industrial distribution of all employed immigrants and non-immigrants in 2010 based on the Labour Force Survey. The service industry is the major employer of immigrants and non-immigrants in Atlantic Canada, within which education and health and social assistance sectors are major employers of immigrant workers, each employing more than 10 %. These sectors also employ large numbers of recent immigrants. About 12 % of recent immigrants were also employed in the food and accommodation sector, a much higher rate than for overall immigrants.

The distribution of non-immigrants is more broad based. Their employment exceeds 10 % of non-immigrants' total employment only in the retail trade and health and social assistance sectors.

The manufacturing sector hires less than 10 % of non-immigrant and immigrant workers in Atlantic Canada, but close to 10 % of non-immigrants are represented. This is an important finding because manufacturing is affected more by business cycle fluctuations than is any other sector of the economy. One implication of this may be that Atlantic Canada is less vulnerable to the current economic downturn than is central Canada, where the manufacturing sector is larger, hiring about 14 % of that region's total labour force. Another implication is that immigrants in Atlantic Canada are less likely to be affected by the current economic downturn than are non-immigrants.

Table 3 shows the occupational distribution of the immigrant and non-immigrant labour force in Atlantic Canada. Given that the service industry is a major employer, the highest percentages are found in sales and service occupations. Also, higher percentages of immigrants than non-immigrants work as scientists and health professionals and in management. That immigrant occupations are more broad based than those of non-immigrants probably reflects deliberate attempts at government and community levels to attract immigrant workers to fill labour shortages in specific occupations where it is hard to find local workers, perhaps due to either out-migration from the region or the demographic trends in population discussed in Chap. 2.

Table 2 Industrial distribution of employed labour force, immigrants and non-immigrants, Atlantic Canada, 2010

	Non-immigrants	Immigrants	Recent immigrants
Employed labour force	1044.8	41.9	7.8
Goods-producing sector (A)	219.6	6.3	1.2
Agriculture	15.5	1	0
Forestry, fishing, mining, oil and gas	39.9	0.6	0
Utilities	10.4	0	0
Construction	78.6	1.8	0
Manufacturing	75.2	2.7	0.6
Services-producing sector (B)	825.2	35.6	6.6
Trade	173.4	5.4	0.7
Transportation and warehousing	49.2	1.7	0
Finance, insurance, real estate and leasing	47.9	1.9	0
Professional, scientific and technical services	44	4	1
Business, building and other support services	46.7	2	0.6
Educational services	77.1	4.4	0.8
Health care and social assistance	155.6	6.9	0.8
Information, culture and recreation	40	1.7	0
Accommodation and food services	63.8	3.5	1.2
Other services	48	1.7	0
Public administration	79.4	2.5	0

Source and notes Statistics Canada Labour Force Survey, 2010, special tabulations. Industrial definitions are based on the North American Industry Classification System 2002. Totals may not agree due to industry not being stated or suppression of data due to small numbers. Zeros could mean no employment or too few observations to report

Table 3 Occupational distributions of non-immigrants and immigrants, Atlantic Canada, 2010

Occupation and labour force	Non-immigrants	All immigrants	Recent immigrants
Total labour force	1168.7	45.4	9.1
Employed labour force—all occupations	1044.8	41.9	7.8
A Management	92.6	5.6	0.7
B Business, finance and administrative	180	6.2	1.5
C Natural and applied sciences and related	60.5	5.0	1.2
D Health	85.8	4.3	0.5
E Social science, education, government service and religion	89.6	4.8	0.7
F Arts, culture, recreation and sport	22.4	1.2	0
G Sales and service	267.6	8.2	1.6
H Trades, transport and equipment operators and related	164.5	3.9	0.7
I Occupations unique to primary industry	46.4	1.2	0
J Occupations unique to processing, manufacturing and utilities	35.3	1.4	0

Source and notes Based on Statistics Canada, 2010 Labour Force Survey, special tabulations. Occupation definitions are based on 2006 National Occupational Classification for Statistics (NOCS)

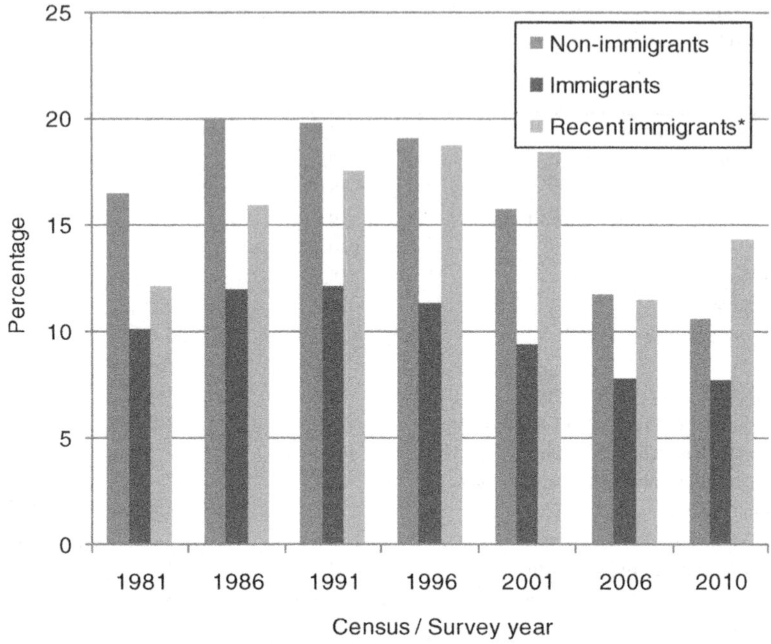

Chart 3 Unemployment Rates among Immigrants and Non-Immigrants in Atlantic Canada, 1981–2010. *Source and note* Table A15. *Those who arrived within the past 5 years of a census/survey date. The labour force activity data based on a census are reported for the year before the census year

Unemployment Rates

Unemployment rates among immigrants and non-immigrants are shown in Chart 3 for the period 1981–2010, which shows that once in the labour force, immigrants do better overall in terms of finding a job than do non-immigrants. The unemployment rates are consistently lower among immigrants and have fallen since 1991, but so have these rates among non-immigrants.

Recent immigrants, however, experienced higher unemployment rates than their counterparts as reported in previous censuses. In 1981 and 1986, the unemployment rate among recent immigrants was 4 % lower than among non-immigrants. However, the gap has narrowed over time, and in 2001, unemployment among recent immigrants was 2.5 % higher than among non-immigrants. Recent immigrants again did slightly better than non-immigrants in terms of finding a job in 2006, but those who arrived later had greater difficulty.

That unemployment rates fell among both immigrant and non-immigrant populations toward the end of the 1981–2001 period indicated that the higher rates among recent immigrants noted in the 2001 census and the 2010 labour force survey cannot be attributed only to the region's prevailing economic conditions at that time. Possible employment barriers faced by recent immigrants due to their lack of knowledge of official languages (English or French), lack of credential recognition, or possible employer discrimination should be investigated in a separate research study.

A remarkable drop in the unemployment rate is observed for all population groups in the post –2006 period, which could be the result of out-migration of their working age members mostly to Alberta, whose booming economy has attracted a working age population from other regions, including Atlantic Canada, thereby reducing the number of people looking for jobs in Atlantic Canada. The drop in the unemployment rate of recent immigrants could also result from their immigration to the region being mostly job oriented, as many came under one of the Provincial Nominee Programs to fill specific jobs. Zietsma (2007) discusses that in 2006, the employment rate of those immigrants who had arrived 5–10 years before that exceeded the rate of non-immigrants—83.6 % for immigrants and 76.4 % for non-immigrants.

Labour Market Earnings

Chart 4 compares the labour market earnings of both an average immigrant and a recent immigrant with those of an average non-immigrant in the year before each census and in the labour force survey of 2010. Actual dollar amounts are not comparable over time because they may be affected by inflation. Instead, it is more meaningful to analyze the earning gap between groups within a given census year and also the changes in this gap between census years. To facilitate this analysis, the ratio of an average immigrant's earnings to a non-immigrant's earnings has been plotted. A ratio of less than one means that an average immigrant's earnings were lower than those of a non-immigrant; greater than one means the opposite.

Average employment earnings were higher among immigrants than among non-immigrants throughout the 1981–2010 period. However, average earnings among recent immigrants, relative to those of a non-immigrant, fluctuated until 2001 and then stabilized. In fact, the earnings gap between these two groups was narrower in 2010. In short, a recent immigrant in 1981 earned 10 % more than an average non-immigrant in Atlantic Canada but 10 % less than a non-immigrant in 2010.

When comparing an immigrant's earnings with those of a non-immigrant, analysts often control for differences in their demographic and labour market characteristics, such as age, gender, experience and education, which are major determinants of earning differences between individuals. They also examine changes in immigrant earnings relative to immigrants' length of stay in Canada to determine how long it takes for an immigrant in Canada to earn the same income as a non-immigrant. This analysis helps explain the labour market integration of immigrants. A detailed analysis of this issue can be conducted in a separate research study by using micro-data based on the labour force surveys conducted since 2006.

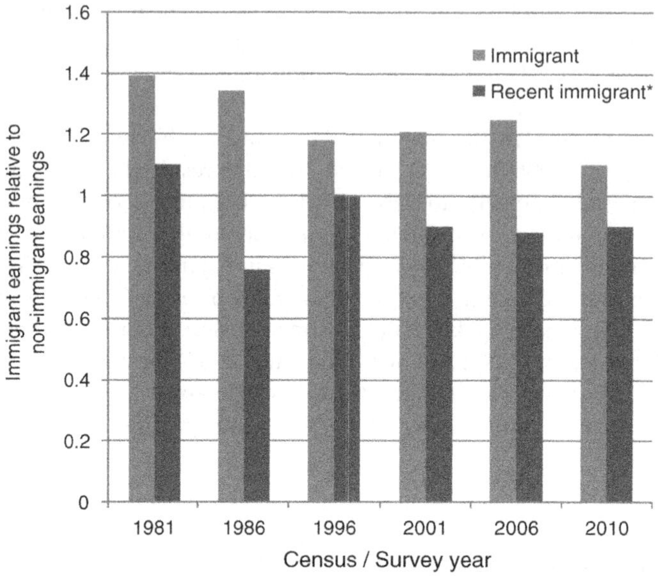

Chart 4 Earnings of an average immigrant relative to those of an average non-immigrant resident in Atlantic Canada, 1981–2010. *Source and note* Table A15. *Those who arrived within 5 years of the census/survey date. Data on recent immigrants in the 1991 census were not available in the census PUMF for the Atlantic provinces. Incomes based on the census are reported for the year before the census year. A ratio of "1" means earnings of immigrants and non-immigrants are equal

Only some broad patterns are presented below on the basis of the 2010 labour force survey, which provides the latest labour market data by immigrant status.

First, we provide a comparative analysis of the earnings of an average immigrant and non-immigrant by controlling for differences in their ages. Chart 5 provides the age-earnings profile of an immigrant versus that of a non-immigrant resident of Atlantic Canada aged 15 years and over based on labour market earnings in 2010. These profiles are based on the earnings of different individuals in the survey, thus assuming that incomes of different individuals at different points in their life cycles can represent the incomes of one typical individual at those same points in his/her life cycle. However, this assumption may be challenged on the grounds that immigrants come from diverse backgrounds and face different labour market challenges. Differences in their educational attainment levels and employers' perceptions of the quality of their education may also have correspondingly different effects on different immigrant groups, so the age-earnings profiles plotted in Chart 5 should be interpreted with some caution.

Despite the above caveats, the shapes of the earning curves in Chart 5 are as expected. Growth in earnings is faster when an individual is young because he/she accumulates more human capital (such as training and experience) when young than when older. Earnings then decline at retirement. However, it is observed that

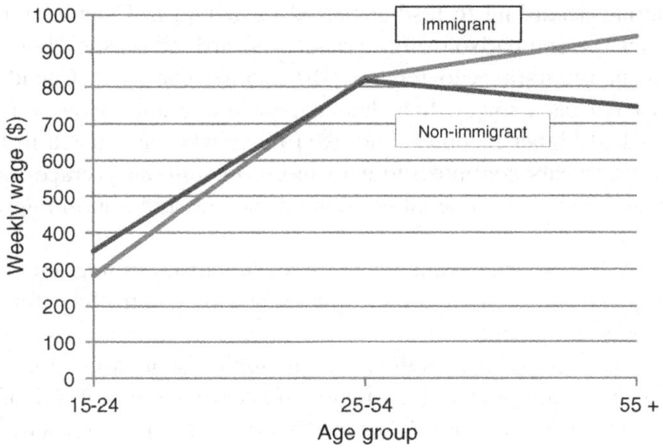

Chart 5 Age-earnings profile of an immigrant and a non-immigrant resident of Atlantic Canada, 2010. *Source* Table A17

an average immigrant's earnings rise faster than and exceed those of a non-immigrant after age 25. Even after reaching the retirement age, the immigrant earns more. Over his/her life cycle, then, an average immigrant in Atlantic Canada earns higher employment income than a non-immigrant. Thus, the average earnings results for 2010 as reported earlier in Chart 4 (above) may be viewed as stable.

Immigrants' earnings are also affected by their length of stay in Canada. A newcomer lacks Canadian labour market experience and information about availability of jobs and workplace culture, factors that may result in an under-utilization of skills, thereby adversely affecting labour market performance and possibly resulting in lower earnings. Chart 6 plots the 2010 labour market earnings

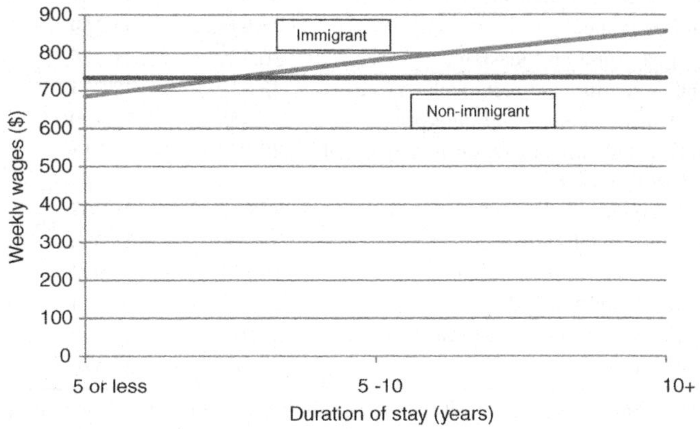

Chart 6 Average earnings of an immigrant by duration of stay and of a non-immigrant in Atlantic Canada, 2010. *Source* Table A18

of an immigrant according to his/her period of arrival in Canada and those of a non-immigrant for comparison. Immigrants who arrived during 2006–2010 were the most recent immigrants to report 2010 income earned in Canada, which is observed to have been lower than that of non-immigrants. However, all earlier entry cohorts had higher incomes, and even those who had stayed in the country for more than 45 years continued to earn more. In sum, an average immigrant in Atlantic Canada earns the same labour market income as a non-immigrant 5 years after arrival.

When more data become available, incomes of various immigrant entry cohorts can be compared in a multivariate framework by controlling for education, ethnicity, gender and country of origin.

Table 4 (next page) compares the average employment earnings of immigrants with those of non-immigrants in 20 major industrial sectors of Atlantic Canada. Overall, an average immigrant worker in Atlantic Canada earns about 27 % more than a non-immigrant, but a recent immigrant earns about 4 % less, perhaps due to

Table 4 Average employment incomes of non-immigrants and immigrants by industry, Atlantic Canada, 2005

Industry	Non-immigrants ($)	All immigrants ($)	Recent immigrants ($)
All industries	32,314	41,081	31,019
1 Agriculture, forestry, fishing and hunting	22,338	25,222	14,220
2 Mining and oil and gas extraction	61,236	79,777	65,590
3 Utilities	58,404	77,045	–
4 Construction	30,642	34,951	20,936
5 Manufacturing	34,843	38,821	37,034
6 Wholesale trade	38,807	41,515	27,616
7 Retail trade	21,044	27,037	12,945
8 Transportation and warehousing	36,138	37,024	22,435
9 Information and cultural industries	39,882	41,181	20,743
10 Finance and insurance	45,948	44,269	15,079
11 Real estate and rental and leasing	30,309	31,356	–
12 Professional, scientific and technical services	43,319	48,324	38,301
13 Management of companies and enterprises	56,055	72,700	26,821
14 Administrative and support, waste management and remediation services	20,042	21,718	14,689
15 Educational services	38,083	48,767	34,531
16 Health care and social assistance	34,700	61,036	67,749
17 Arts, entertainment and recreation	19,846	19,033	9,193
18 Accommodation and food services	13,920	19,814	15,676
19 Other services (except public administration)	22,962	23,749	19,842
20 Public administration	47,943	57,378	40,813

Source and notes Statistics Canada, 2006 Census of Population, special tabulations. Provincial data were used to compute weighted averages. Data are for all employed persons (with and without income) in each industry. "–" means data not available for confidentiality reasons (posted as zero in the B20/B20 file supplied by Statistics Canada) his/her lack of experience in Canadian labour markets. On average, an immigrant's income is higher than that of a non-immigrant in 18 of the 20 industries reported, with the highest income difference being in health care and social assistance industries, where an immigrant earns about 76 % more than a non-immigrant. Mining and oil and gas extraction industries offer the highest employment incomes. In these industries, even an average recent immigrant is paid more, overall, 30 % more than a non-immigrant. In the two industries in which immigrants are paid less than non-immigrants—finance and real estate and arts, culture and recreation—the immigrant income disadvantage is at 4 % or less.

The income difference between the two populations is largely determined by differences in their age distributions and human capital (education and experience). Chart 5, which plotted the age-earnings profiles of an average immigrant and non-immigrant, showed that an average immigrant in the 25 and over age group, earned more than a non-immigrant in that group. A more complete analysis should also obtain age-earnings profiles of an immigrant and a non-immigrant at different education levels. Impacts of other factors, such as knowledge of English and/or French, should also be investigated in an econometric model.

Immigrants' Home Country Educational Credentials and Labour Force Activity

One important issue in the labour market outcome of immigrants is credential recognition. It is argued that immigrants coming from certain countries, especially those coming from less developed countries, face employment barriers because the education and employment experiences acquired in their countries of origin are not recognized in Canada. This lack of immigrant credential recognition has at least three consequences: (1) these immigrants could face higher unemployment rates; (2) even if employed, these immigrants may be working in a job that does not suit their qualifications; and (3) Atlantic Canada's economy suffers a loss by not fully benefiting from the human capital of its new residents.

To investigate foreign credential recognition, we used available data from the 2006 census to review some indicators of the labour market performance of those who finished their post-secondary education in a different country. These indicators are reported for the total population (regardless of immigrant status) in Table 5. Results showed that those who obtained their post-secondary education outside, rather than inside, Canada generally have lower labour force participation and employment rates. Unemployment rates are also generally lower among foreign degree holders. However, data analyzed by the countries where education was obtained produced some mixed results. Those who finished their post-secondary education in India, Pakistan and the Republic of Korea have higher unemployment rates; all others have lower rates, regardless whether they obtained their education in an English-speaking country.

Table 5 Labour force activity of the total population by location of post-secondary education, Atlantic Canada, 2005

Location of study (1)	Participation rate (%) (1)	Employment rate (%) (2)	Unemployment rate (%) (3)	Labour force, number (4)
Overall	62.57	55.37	11.52	1,182,970
No post-secondary certificate, diploma/degree	51.85	43.72	15.67	518,970
Post-secondary certificate, diploma/degree	74.64	68.47	8.27	664,005
Inside Canada	75.30	69.02	8.33	639,120
Outside Canada	60.90	56.88	6.61	24,885
United States of America	62.77	59.09	5.82	10,225
United Kingdom	52.34	49.92	4.74	4,645
India	63.39	58.48	9.86	710
Philippines	78.30	73.58	2.41	415
China, People's Republic of	65.04	60.18	6.12	735
Germany	57.06	53.74	6.31	1,030
France	73.55	69.03	7.02	570
Poland	72.27	73.95	2.33	430
Pakistan	73.13	64.18	8.16	245
Korea, South (Republic of)	36.14	31.33	13.33	150
Other	63.70	57.57	9.53	5,720

Source Author's calculations using provincial data based on Statistics Canada—2006 Census. Catalogue Number 97-560-XCB2006025. For detailed notes, see the publication on www.statcan.gc.ca

Notes

1. Place where the highest post-secondary certificate, diploma or degree was completed
2. Percentage of population 15 and over in the labour force
3. Percentage of population 15 and over employed
4. Percentage of labour force unemployed

Skilled and Business Immigrants in the Atlantic Economy

Skilled workers and business immigrants make up the so-called "economic" immigrant class. CIC defines the skilled worker class of immigrants as "people who may become permanent residents because they are able to become economically established in Canada" (www.cic.gc.ca).

"Business immigrants are people who can invest in, or start, businesses in Canada and are expected to support the development of a strong and prosperous Canadian economy. The Business Immigration Programs seeks to attract to Canada people experienced in business. Business immigrants are selected based on their ability to become economically established in Canada" (www.cic.gc.ca).

CIC also divides business immigrants into three classes: investor, entrepreneur and self-employed. Investors are experienced persons who must demonstrate business experience, have a minimum net worth of $800,000, and make an investment of $400,000. Entrepreneurs are experienced persons who will own and actively manage a business in Canada that will contribute to the economy and create jobs. Entrepreneurs must have business experience and a minimum net worth of $300,000 and are subject to conditions on arrival in Canada that include creation of at least one job for Canadian citizens. Finally, the self-employed must intend and be able to create their own employment by operating a business or farm in Canada and are also expected to contribute to Canada's cultural or athletic life.

Immigration policy falls under federal jurisdiction. While each Atlantic province abides by national objectives that allow immigration on humanitarian grounds for refugees and others for family re-unification, the main need is to foster economic immigration suited to each province's long-term needs. Provinces can and do work out special agreements with the federal government to promote immigration perceived to be in their interest. For example, Quebec has had such an arrangement for many years, and smaller provinces such as Saskatchewan, Manitoba and all four Atlantic provinces have engaged in such initiatives since the late 1990s. The PNP reflects this new thinking of enhancing the positive impact of immigration in each province.

Since economic immigration is likely to be central to policy in the Atlantic region, it is important to examine it in detail. Therefore, this section analyzes data from a number of sources to shed light on the various dimensions of economic immigration to Atlantic Canada. Our discussion will be divided into two parts—one focusing on skilled workers and the other on business immigrants.

While PRDS provides data on the arrival of immigrants under the skilled and business classes, no direct data are available on the economic performance of those who arrived under these classes to compare with data on non-immigrants as a reference group. However, the economic performance of immigrants who practiced different occupations in Canada can be assessed using census information. Since the present part of this study focuses on those immigrants who can establish themselves economically in Canada, census data are analyzed only for those who worked as professionals and managers. Human Resources and Skill Development (HRSD) lists these two occupations ("O" and "A") among the top in terms of educational and skill requirements. Finally, to assess the performance of immigrants in the business sector, data are analyzed for those who declared themselves "self-employed" on the census questionnaire.

Immigration of Highly Skilled Workers to Atlantic Canada

Chart 7 shows the trends of three occupational groups of immigrants destined for Atlantic Canada's labour force: "highly skilled," "medium skilled," and "low skilled" immigrants, groupings based on the occupational classifications as defined by Human Resource and Social Development Canada (HRSDC).[1] Note here,

[1] http://www5.hrsdc.gc.ca/noc/english/noc/2011/Welcome.aspx

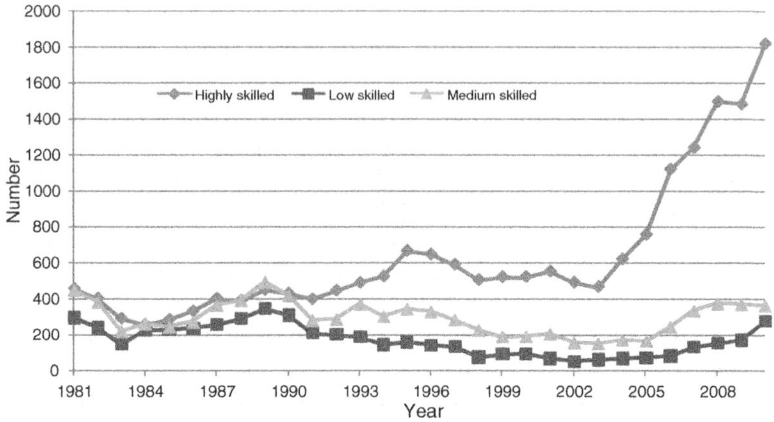

Chart 7 Immigrants destined to Atlantic Canada by skill classifications, 1981–2010. *Source and note* Table A19. National Occupational Classifications (NOC) were further classified as highly skilled = "O" and "A"; medium skilled = "B", low skilled = "C" and "D"

however, that only about half the immigrants destined to work in Atlantic Canada have their occupations identified.

Throughout the period 1981–2010, highly skilled immigrants (i.e., groups "O and "A", managers and professionals) comprised a significant proportion of those destined for the labour force of Atlantic Canada. Since the early 1990s, their levels have exceeded those of medium- and low-skilled immigrants, and after some decline between 1994 and 2002, their numbers have risen sharply, reflecting the policy emphasis of attracting highly skilled immigrants.

Table 6 captures the trend of highly skilled immigrant inflows by showing the distribution of one sub-group of resident highly skilled immigrants, i.e., professionals, across the 1991, 1996, 2001 and 2006 census years in selected occupations. The numbers of non-immigrant professionals are also provided for comparison. (These data are based on a 20 % sample of individuals. Statistics Canada cautions that some numbers in Table 6 may not be accurate due to some miscoding of occupations. Further, in the 2006 census, occupations were coded using a different scheme, so some categories within an occupational group had to be re-grouped for the present study to make them comparable across censuses. However, Table 6 data do permit general comparisons between immigrants and non-immigrants and also between these groups and recent immigrants.)

Overall, immigrants constitute a significant proportion of the selected professionals in Atlantic Canada, as shown in Table 6. However, due to declining immigrant inflows, the number of non-immigrant professionals grew at a faster rate during 1991–2001, with the strongest growth coming in the latter half of that period. Growth in the number of non-immigrant professionals was more broad-based during this period than that of immigrants. Since then, the increase in immigrant professionals has been faster because many came to fill specific jobs

Table 6 Professionals in Atlantic Canada by selected occupations: immigrants and non-immigrants, 1991–2006 censuses

Professionals by occupation	Census year			
	1991	1996	2001	2006
Business and finance				
Auditors, accountants and other investment professionals				
Non-immigrants	6,610	7,170	11,395	12,915
Immigrants	370	370	485	725
Recent immigrants	10	10	40	120
Other professionals in business and finance*				
Non-immigrants	1,865	2,275	2,985	5,295
Immigrants	90	125	245	250
Recent immigrants	5	0	10	20
Natural and applied science and related				
Engineers				
Non-immigrants	6,465	5,960	6,520	7,825
Immigrants	970	640	870	925
Recent immigrants	135	110	100	230
Computer and information systems				
Non-immigrants	6,315	6,870	14,675	10,505
Immigrants	540	650	1065	940
Recent immigrants	85	100	200	270
Other scientists*				
Non-immigrants	3,815	3,860	4,375	4,735
Immigrants	610	445	585	680
Recent immigrants	80	25	85	115
*Health professionals***				
Physicians, dentists, and veterinarians				
Non-immigrants	3,185	3,660	4,080	4,805
Immigrants	1,455	1,290	1,420	1,845
Recent immigrants	295	220	145	290
Other health professionals*				
Non-immigrants	3,165	3,900	4,980	6,145
Immigrants	295	355	360	350
Recent immigrants	25	55	20	50
*Social science, education, government services and religion**				
Judges, lawyers, Quebec notaries				
Non-immigrants	2,665	2,910	3,155	3,960
Immigrants	140	175	215	235
Recent immigrants	20	0	10	10

(continued)

Table 6 (continued)

Professionals by occupation	Census year			
	1991	1996	2001	2006
Teachers and professors				
Non-immigrants	40,490	40,080	37,525	43,050
Immigrants	3,795	3,765	3,485	3,730
Recent immigrants	255	385	240	570
Other professionals in social science, education, government services and religion*				
Non-immigrants	21,105	22,350	24,440	22,225*
Immigrants	1,445	1,540	1,445	1,475*
Recent immigrants	130	150	140	165*
Arts, culture, recreation, and sports				
Musicians and singers				
Non-immigrants	1,210	1,575	1,495	1,645
Immigrants	215	170	170	215
Recent immigrants	45	25	25	15
Other professionals in arts and culture*				
Non-immigrants	5,340	5,905	6,870	8,090
Immigrants	775	805	750	845
Recent immigrants	100	35	75	85

*Computed total of all other professionals. **Excludes nurse supervisors and registered nurses
Source 1991–2001 data are from the Table entitled: "Occupation—1991 Standard Occupational Classification (Historical) (707B), Selected Labour Force, Demographic, Cultural, Educational and Income Characteristics (252) and Sex (3) for Population 15 Years and Over, for Canada, Provinces, Territories and Census Metropolitan Areas 1, 1991–2001 Censuses—20 % Sample Data". Statistics Canada. Catalogue number 97F0012XCB2001048.Data for 2006 are from Table entitled: "Occupation—National Occupational Classification for Statistics 2006 (720C), Sex (3) and Selected Demographic, Cultural, Labour Force, Educational and Income Characteristics (273) for the Population 15 Years and Over of Canada, Provinces, Territories, Census Metropolitan Areas and Census Agglomerations, 2006 Census—20 % Sample Data". Statistics Canada. Catalogue number 97-564-XWE2006005

under PNPs in all provinces. Except for two occupational groups, population increased within each demographic group.

Of particular importance is the increased number of immigrant scientists and engineers, who are generally viewed as direct contributors to economic growth by promoting innovation. The dramatic rise in immigrant health professionals is also worth noting. Atlantic Canada has faced physician shortages in recent years as many moved out of the region, a shortage felt especially in rural areas, where many hospitals have had to close their emergency department one or more days each week because of the unavailability of professionals. To overcome this problem, provinces have taken special measures to attract internationally trained physicians. (See Box 3 in Chap. 3 for the measures adopted by Nova Scotia and

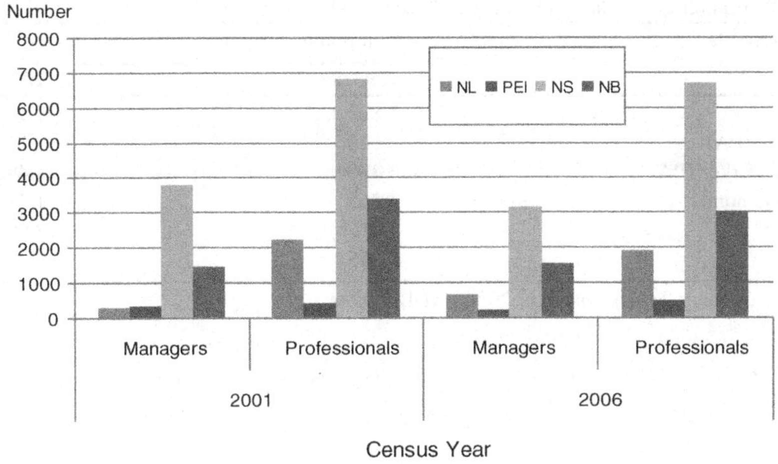

Chart 8 Provincial distribution of highly skilled immigrants, Atlantic Canada, 2001 and 2006. *Source* Table A20

Newfoundland and Labrador.) Also, the number of university teachers and professors who came during 2001–2006 was more than twice the number who arrived during the previous five-year period.

The 2006 census data in Table 6 also revealed that about 87 % of immigrant professionals in Atlantic Canada worked in the service sector within which, 33 % were working in the education sub-sector and 25 % in health and social assistance. Public administration accounted for 8 %, manufacturing for 2 %, and agriculture, mining and utilities for 1.7 %.

Provincial Distribution of Highly Skilled Immigrants

Most highly skilled immigrants in Atlantic Canada live in Nova Scotia (Chart 8), more than twice as many as in New Brunswick. Relative to managers, the number of professionals living in Newfoundland and Labrador is the highest in the region, while Prince Edward Island has an even distribution of managers and professionals.

Business Immigration

As noted in Chap. 2, business immigration to Atlantic Canada declined dramatically over the 1998–2008 period, from close to 300 in 1998 to only 13 in 2008. Also noted, the increased inflow of business immigrants during the mid- to late 1990s was an anomaly. The region also attracted business immigrants who initially

Table 7 Immigrant tax filers in Atlantic Canada, 2008 taxation year

Immigrant class	Immigrants arriving	
	1993–2007	2003–2007
Family	4,825	2,155
Business (PA)	620	115
Skilled worker (PA)	3,480	1,495
Other economic	4,265	1,395
Refugee	4,115	3,090
Other	1,055	550

Source Summary Tables based on 2008 IMDB Tables

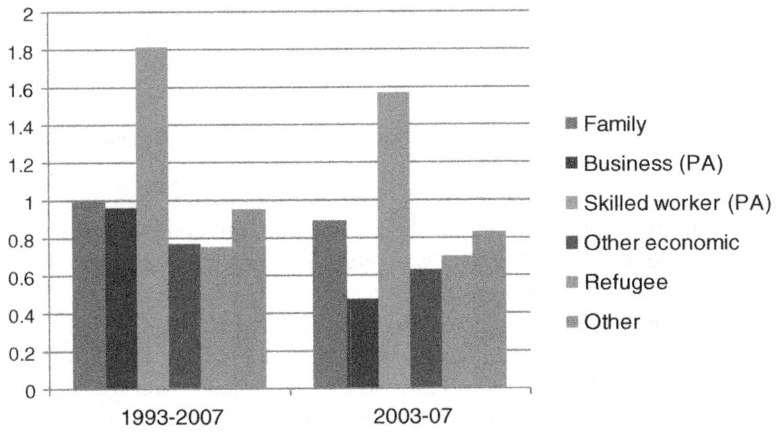

Chart 9 Labour market earnings of an average immigrant by class relative to those of an average tax filer, Atlantic Canada, 2008. *Source* Table A21 and A22

landed in other regions of Canada (Table 6 in Chap. 3). No information is yet available on the industrial distribution, occupation or countries of origin of those immigrants who arrived in Canada as business immigrants. However, the IMDB data do permit an analysis of their earnings in Canada according to their period of immigration. Table 7 reports the numbers of immigrant tax filers by six immigrant classes grouped by their periods of arrival in taxation year 2008.[2] Three sub-classes of economic immigrants are reported: business (principal applicants only), skilled worker (principal applicants only), and other economic. The economic class comprises the most tax filers, followed by family, refugee and other classes.

Chart 9 plots the labour market earnings (employment and self-employment) of an average immigrant tax filer who landed as part of any of the six immigrant classes relative to the earnings of an average tax filer resident in Atlantic Canada. (IMDB does not include data for non-immigrants, so the average earnings reported

[2] These were the latest data available at the time of writing this book.

for all tax filers on the Canada Revenue Agency website were used for comparison.) Separate plots were obtained for those who landed in Canada during 1993–2007 and during 2003–2007, filed their income tax returns in 2008, and reported earning employment and self-employment incomes in 2008.

Chart 9 results show that a family class immigrant resident of Atlantic Canada who had been in the country for 1–15 years earned the same as an average resident tax filer in Atlantic Canada in 2008, while a skilled worker class immigrant earned more than 80 % of that of the average tax filer. All other immigrant classes experienced an earnings disadvantage relative to the average tax filer, with refugees experiencing the greatest disadvantage. Recently arrived (2003–2007) skilled workers also experienced an earnings advantage close to 60 % over those of the average tax filer (data are not available for non-immigrant tax filers). All other recently arrived immigrants experienced an earnings disadvantage in 2008, with business class immigrants showing the poorest performance; however, this could be attributed to the failure of the business immigration program in Nova Scotia, the region's largest province.

A summary profile of immigrant entrepreneurs in Atlantic Canada based on survey reports prepared for different government departments is provided in the Box 1. Most immigrant entrepreneurs are in Nova Scotia, followed by those in New Brunswick, Newfoundland and Labrador, and Prince Edward Island. These entrepreneurs may or may not have entered Canada as business class immigrants, and many also migrated from outside Atlantic Canada.

1 Immigrant Entrepreneurs in Atlantic Canada

Newfoundland and Labrador

Newfoundland and Labrador (2007b) profiles 20 immigrant business entrepreneurs living in that province. Nine were engaged in the hospitality industry, managing businesses such as restaurants, pizza making and delivery, bakeries and catering. Most had come from Europe, but one business woman from Pakistan ran a restaurant, and one businessman from Lebanon ran a pizza shop. The remaining eleven ran various businesses in arts and culture, trading and apparel, as well as a dental clinic, an electronic products shop, an imported furniture store, a sports shop, a paint shop, and a skin care store. Most had arrived since the 1990s, some landing directly in the province and some moving from other provinces.

Prince Edward Island

The Population Secretariat of Prince Edward Island commissioned a testimonial profiling immigrant entrepreneurs living on the island (Baldacchino and Fall 2008) Of 60 businesses identified for the study, 47 were located in Queens County. Most entrepreneurs were involved in catering, including food outlets and ethnic restaurants (21 firms), followed by those in accommodation, including beach cabins, bed and breakfasts, cottages, hotels and campgrounds (13 firms); craft, professional, consulting and other services (12 firms); farm, food, beverage and natural products (nine firms); and computer gaming, automation, aerospace and navigation instruments (five firms). Although immigrant entrepreneurs have been on the Island since 1977, most sampled in this study had come from Europe since the late 1990s, with some arriving from Asia and Africa.

Nova Scotia

In 2008, the Metropolitan Immigrant Settlement Agency (MISA) surveyed a sample of 51 immigrant entrepreneurs living in the province to determine the factors that facilitate or hinder their success (Sharif 2009). Most have come to the province since 1990, mainly under the family or skilled worker category rather than as entrepreneurs. Europe is the dominant source region of these immigrants, followed by the Middle East and Asia. Immigrants from Africa, North America (the USA) and Central America make up the rest. Most live within the boundaries of Halifax Regional Municipality, and their businesses are likely to be in food, import or retail. As will be discussed later, the province has now terminated its business immigration program.

New Brunswick

No published report profiles immigrant business entrepreneurs in New Brunswick. Information provided by the Population Growth Secretariat (Immigration Division) indicates that business immigrants have come from a variety of countries, the top three (in order of ranking) being China, South Korea and Iran. Some business entrepreneurs have taken over existing smaller retail, service or accommodation businesses or launched new ventures in these fields and others, but many are still in the adjustment and planning stages.

2 The Immigrant Investor Program

Canada has had some form of Immigrant Investor Program (IIP) since 1986. IIP seeks to promote economic growth in all regions of Canada by attracting experienced business persons and new investment capital. Before 1999, each province administered its own funds, competing aggressively with other provinces for available capital. Between 1986 and 2002, the total amount of capital attracted was $8.7 billion from about 27,500 investors. Factoring in the dependents of these applicants means that IIP has attracted about 82,500 people to Canada.

IIP was modified in 1999 to tighten eligibility criteria and bring the flow of capital into a single Canada Fund administered by CIC. This replaced the provincially administered funds except in Quebec, which still runs its own fund independently of the Canada Fund. The other provinces can sign agreements with CIC to participate in, and receive allocations from, the Canada Fund. CIC allocates half the available funds, equally divided, to approved provincial funds; the other half is distributed according to provincial shares of the gross domestic product (GDP).

Potential immigrant investors can approach, or be recruited by, financial institutions that have signed agreements with CIC to act as facilitators for the application process. These institutions are all members of the Canada Deposit Insurance Corporation. Currently, 11 institutions carry out this agency role, including most of Canada's major banks. Each receives a fee of $28,000 for each immigrant investor on issuance of a visa fee paid from the $400,000 an investor contributes to the fund.

Eligibility criteria now state that an applicant must pay $400,000 to the Receiver General of Canada, have a net worth of at least $800,000, and demonstrate the successful operation, management and control of a business. Previous criteria were more modest, with a tier system that set required investment levels higher for the three biggest provinces, a system that made it easier for smaller provinces to attract immigrant investors. Under the IIP, the principal ($400,000) is repaid to investors about 5 years after payment into the fund. The repayment is guaranteed by the participating province, which effectively means that the province also pays the agent's fee ($28,000) to attract the investor in the first place.

Most parts of Canada have benefited from the flow of funds under IIP since the program's inception but especially Quebec, which attracted 62 % of the total capital that flowed into Canada between 1996 and 2005. In Atlantic Canada, Prince Edward Island did particularly well for its size, attracting almost 4 % of the total capital flows, and Nova Scotia also benefited from the program. Not all provinces have applied for participation in the re-designed fund, however; neither Nova Scotia nor New Brunswick was a participant in 2005.

After IIP was re-designed, there was a dip in capital flows to individual provinces, but these have since recovered. Prince Edward Island received $5.4 million in 2000 but $32 million in 2005. Also, participating provinces have considerable leeway in how they invest the funds they receive, the only constraint being that the money be used for economic development. The funds IIP raises have been used by provinces to undertake investments without placing further burdens on fiscal capacities. In Prince Edward Island, funds that began in the early 1990s have been invested in both public and private projects, including golf courses, resorts and other tourism infrastructure, diversified manufacturing, and commercial real estate. IIP does not require investors to live in the province where their capital is invested, which raises the question, "Is IIP used to attract immigrants or to attract capital?"

Rural–Urban Labour Force Division

As discussed in Chap. 3, about one-third of Atlantic Canadians live in rural areas. Table 8 (above) shows the distribution of the Atlantic labour force by rural–urban areas and also by gender and immigrant status. These data are based on the 2010 Labour Force Survey, which uses the broad classification of urban and rural areas into Census Metropolitan Areas and outside of Census Metropolitan Areas. Overall, immigrants comprise 3.7 % of the total labour force. About 24 % of all immigrants, but only 15 % of recent immigrants who are either working or looking for a job, do so in rural areas. Among the non-immigrant labour force, this percentage is higher at about 39 %.

Rural and urban unemployment rates are similar among immigrants. However, among non-immigrants, the urban unemployment rate is significantly lower than that in rural areas, and the unemployment rate among all immigrants in the labour force is lower than that for non-immigrants in both urban and rural areas. Regional and urban unemployment rates among female immigrants are lower than among males. In urban Atlantic Canada, the unemployment rate of females only slightly exceeds that of non-immigrant females. Among recent immigrants, the unemployment rates exceed those of non-immigrants regionally and in urban areas, possibly indicating their lack of labour market experience. No data are available on

Table 8 Labour force and unemployment rates, immigrants and non-immigrants by rural and urban regions, Atlantic Canada, 2010

	Rural (Non-CMA/CA)			Urban (CMA/CA)			Regional		
	Male	Female	Total	Male	Female	Total	Male	Female	Total
Labour force (thousands)									
Immigrants	5.9	4.9	10.8	19.4	15.2	34.6	25.2	20.1	45.4
Recent immigrants	0.9	0.5	1.4	4.7	3.0	7.7	5.5	3.6	9.1
Non-immigrants	236.7	215.8	452.5	361.5	354.7	716.2	598.2	570.5	1168.7
Unemployment rate (%)									
Immigrants	10.2	NA	7.4	8.2	7.2	7.8	8.7	6.5	7.7
Recent immigrants	NA	NA	NA	14.9	16.7	15.6	14.5	13.9	14.3
Non-immigrants	16.6	11.8	14.3	9.7	6.9	8.3	12.4	8.7	10.6

Source Labour Force Survey of 2010 (special tabulations). NA refers to data not available due to suppression by Statistics Canada

Table 9 Industrial distribution of employed labour force by immigrant status, Urban and Rural Atlantic Canada, 2010

Total	Immigrants			Recent immigrants			Non-immigrants		
	Total 41.9	Urban 31.9	Rural 10	Total 7.8	Urban 6.5	Rural 1.3	Total 1044.8	Urban 657.1	Rural 387.7
Goods-producing sector	6.3	3.8	2.6	1.2	0.7	0.5	219.6	105.4	114.2
Agriculture	1	0	0.8	0	0	0	15.5	3.5	12
Forestry, fishing, mining, oil and gas	0.6	0	0	0	0	0	39.9	12.1	27.7
Utilities	0	0	0	0	0	–	10.4	7.2	3.3
Construction	1.8	1.3	0.5	0	0	0	78.6	45.8	32.7
Manufacturing	2.7	1.5	1.1	0.6	0	0	75.2	36.8	38.5
Services-producing sector	35.6	28.2	7.4	6.6	5.8	0.8	825.2	551.7	273.6
Trade	5.4	3.6	1.8	0.7	0.5	0	173.4	111.7	61.8
Transportation and warehousing	1.7	1.2	0.5	0	0	0	49.2	30.1	19.1
Finance, insurance, real estate and leasing	1.9	1.5	0	0	0	–	47.9	35.6	12.4
Professional, scientific and technical services	4	3.5	0	1	1	0	44	35.7	8.3
Business, building and other support services	2	1.7	0	0.6	0.5	0	46.7	33.5	13.3
Educational services	4.4	3.9	0.5	0.8	0.8	0	77.1	50	27.1
Health care and social assistance	6.9	5.1	1.8	0.8	0.7	0	155.6	97.1	58.5
Information, culture and recreation	1.7	1.5	0	0	0	–	40	29.8	10.2
Accommodation and food services	3.5	2.8	0.6	1.2	1	0	63.8	42.1	21.7
Other services	1.7	1.1	0.5	0	0	0	48	29.2	18.7
Public administration	2.5	2.2	0	0	0	0	79.4	57	22.4

Labour force survey (2010) (Special tabulations). Zeros indicate either no employment or data not available due to suppression by Statistics Canada

the unemployment rates among female immigrants and on recent immigrants in rural areas.

About 85 % of all immigrants work in the services-producing sector, 80 % of whom live in urban areas, with a 60–40 urban–rural distribution in the goods-producing sector. As Table 9 data show, these distributions are similar among recent immigrants; however, 87 % of those working in the services-producing sector are in urban areas. Most non-immigrants employed in the goods-producing sector work in rural areas. In the services-producing sector, their employment is more evenly divided between urban and rural areas than it is for immigrants.

Data on the occupational distribution of immigrants living in rural areas of Atlantic Canada are not available. However, information based on discussions with the Rural Secretariat and some community organizations suggests that most

American and western European immigrants are engaged in various occupations, for example, farming, the hotel industry, retail, and professional occupations. Many involved in farming were attracted to Atlantic Canada because of the highly competitive land prices. Also, recent immigrants from India and China may have come under the PNP aimed at meeting the shortages of professionals in rural areas. Among them, Indian immigrants are located mainly in Newfoundland and Labrador, while Chinese immigrants are found mostly in Nova Scotia. A Statistics Canada study (Beshiri and He 2009) found that the share of Asian immigrants in rural zones of Atlantic Canada equals or surpasses that of eastern and southern Europeans. The study also found that recent immigrants are more mobile than non-immigrants, being more likely to move into and out of rural areas within Atlantic Canada, especially in Nova Scotia and New Brunswick.

Immigrants' Use of Government Transfer Payments

A common public misperception about the economic impact of immigrants is that they rely more on government transfers, such as public assistance income, employment insurance benefits, welfare payments and old age security income. Rather, that immigrants have higher employment incomes suggests that their reliance on government transfer payments would be lower, as confirmed by the data in Chart 10 for the period 1981–2006 for which data are available. In fact,

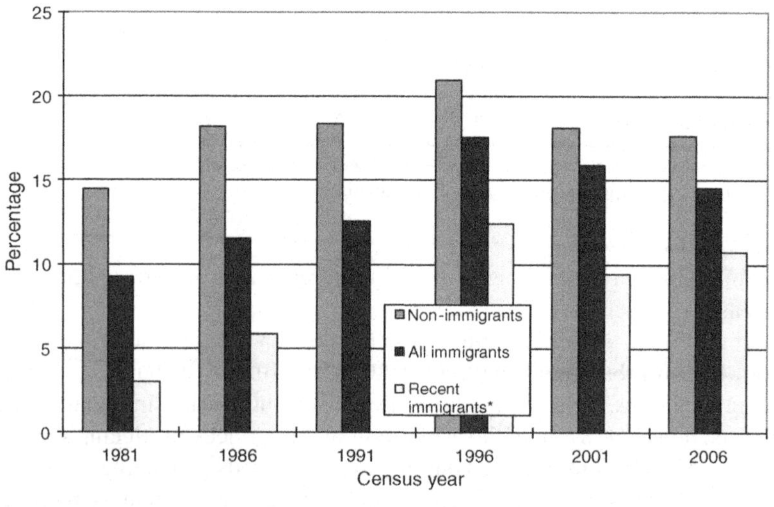

Chart 10 Percentage of total income received as government transfers by immigrants and non-immigrants in Atlantic Canada, 1981–2006. *Source and note* Table A15. *Those who arrived within 5 years of the census year. Data on recent immigrants in the 1991 census were not available in the census PUMF for the Atlantic provinces. All income data are reported for the year before each census year. Data for the post-2006 period are not available

immigrants overall, as well as recent immigrants, received lower percentages of total income as government transfers than did non-immigrants throughout the 1981–2006 period (data are not available for later period).

A major component of government transfer income is age related; for example, both Old Age Security and the Canada/Quebec Pension Plan income are available only to those over 60. Since most immigrants are young at the time of arrival, they do not become eligible for such transfer incomes for a long time after arrival in Canada. Another component of government transfer payments is employment insurance income, which (1) is available only to those who have worked for a certain number of weeks in Canada, (2) supplements the income of the unemployed who contributed to the program while employed, and (3) is determined on the basis of the contribution made by the recipient when employed and by his/her employer. Also, as recent immigrants may not have accumulated sufficient work experience in Canada and may not have contributed enough into the program, many may not be eligible to receive employment insurance.

To summarize, the analysis in this chapter reveals that an average immigrant in Atlantic Canada has strong labour market outcomes. An immigrant who arrived after 2001 faced a lower unemployment rate than one who had arrived earlier. A resident immigrant also earned more income than did a non-immigrant in the post-2001 period. The earnings disadvantage faced by a recent immigrant relative to the earnings of a non-immigrant was the same at the time of the 2006 census as it was in 2001 but narrowed in 2010. An immigrant's income rises faster than that of a non-immigrant over his/her life cycle and exceeds a non-immigrant's income 5 years after arrival.

Important insights can be gained by comparing the above results with those of the rest of Canada. This is not pursued in the present book, but Appendix Table A16 provides a brief comparison with national labour market indicators in 2010. In the case of non-immigrants in Atlantic Canada, their regional unemployment rate is higher than their national rate, while their labour force participation rate and average weekly earnings are lower than the national level participation and weekly earnings for non-immigrants across the country. In the case of all immigrants, their unemployment and labour force participation rates are lower than the corresponding national rates, and their average weekly earnings are higher. In case of recent immigrants, the unemployment rate is lower regionally than it is nationally, while their average weekly earnings are higher. Their labour force participation rate is also higher in the region. These results clearly indicate that immigrants perform better in the labour markets of Atlantic Canada than they do nationally.

As discussed in earlier chapters, new immigrant attraction initiatives in Atlantic Canada largely focus on skilled immigrants. However, skilled immigrants are known to face some difficulty in labour market integration because of the lack of recognition of their human capital acquired in their countries of origin. To overcome this challenge faced in the efficient utilization of immigrant skills, immigration policy makers also emphasize attraction and possible retention of international students. The next chapter presents the outcomes of such initiatives undertaken in Atlantic Canada.

References

Akbari, A. A. (2010). *Labour force participation of immigrants in Nova Scotia Circa: 2006*. Retrieved December 21, 2011, from http://economics.acadiau.ca/tl_files/sites/economics/resources/ACEA/Papers%20and%20Procedings/2010/A.Akbari.2010.pdf.

Baldacchino, G., & Fall, C. (2008). *Immigrant entrepreneurs on Prince Edward Island university of Prince Edward Island* (ISBN 978-0-9810620-0-6).

Beshiri, R., & He, J. (2009). *Immigrants in rural Canada: 2006* (Statistics Canada Cat # 21-006-X).

Newfoundland and Labrador. (2007b). Diversity and opportunity: profiles of immigrant entrepreneurs in Newfoundland and Labrador department of human resources, Labour and Employment (2nd ed.).

Sharif, N. (2009). *Immigrant Entrepreneurs in Nova Scotia: Factors that Contribute to and Hinder Success Atlantic Metropolis Centre Working Paper # 20.*

Zietsma, D. (2007). *The Canadian immigrant labour market in 2006: first results from Canada's labour force survey* (Statistics Canada, Cat # 71-606-XIE2007001).

Chapter 5
International Students in Atlantic Canada

International students are viewed as potential new immigrants. "International graduates are young with advanced English language skills, with fully recognized qualifications, locally relevant professional training and a high degree of acculturation" (Hawthorne 2005). These characteristics are believed to facilitate their integration into both the labour market and society as a whole. In addition, international students are also sources of revenue for local educational institutions and help strengthen local markets for goods and services. Siddiq et al. (2010) found that international students contributed $175 million of new money to the Atlantic Canadian economy in 2009–2010 (see Box 1).

On average, according to data presented by Martha Justus of CIC at the eleventh International Metropolis Conference in Lisbon (4 October 2006), 15–20 % of international students can be expected to eventually settle and work in Canada. In 2007 alone, international students who converted their status into that of permanent residents comprised 15.7 % of such temporary residents in Canada and 27.6 % in Atlantic Canada (CIC 2007). Siddiq et al. (2010) also report the findings of their survey of international students showing that 40 % planned to apply for permanent resident status in Canada. These students thought highly of Atlantic Canada and its educational institutions and were acclimatized to the region.

Attracting international students and retaining them after graduation is one goal of immigration strategies adopted by provincial governments in Atlantic Canada. For example, the Provincial Nominee Program of Nova Scotia includes the component "International Graduate Stream," which fast tracks the landing process for those international students who wish to stay in the province after finishing their studies. The immigration strategy of the province of Newfoundland and Labrador explicitly supports post-secondary institutions and K-12 schools in attracting increasing numbers of international students, with Memorial University being a key partner. New Brunswick has also signed an agreement with the federal government that makes it easier for foreign graduates there to gain an additional year of work experience in their field of study. This change is expected to help graduates who wish to apply for permanent resident status as skilled workers by providing them with additional points on the selection grid in the area of

S. A. H. Akbari, *Immigrants in Regional Labour Markets of Host Nations*, SpringerBriefs in Population Studies, DOI: 10.1007/978-94-007-6244-2_5,

experience. Prince Edward Island also participates in the "International Graduate Stream," one of the four streams under the Labour Impact Category of its revamped PNP. This stream is employer driven and provides the opportunity for Prince Edward Island employers to hire a recent graduate from a recognized accredited Canadian university or college. Universities and colleges in all four provinces are now more aggressively marketing their programs to attract international students.

1 International Students Contribute Significantly to the Atlantic Economy

International students contributed $565 million to the Atlantic Canadian economy in 2009–2010, according to a new study, *The Economic Impact of Post-Secondary International Students in Atlantic Canada*, released today by the Council of Atlantic Ministers of Education and Training (CAMET). In addition to their significant contributions to the region's economy, the study highlights that international students are a major source of potential immigrants for Atlantic Canada.

International students provide an immediate economic impact and return on investment in Atlantic Canada. The study determined that international students contributed $175 million of new money to the region's economy in 2009–2010. Furthermore, international students spent $2.68 of new money in Atlantic Canada for every dollar spent by the four Atlantic governments.

"The findings of this study support Newfoundland and Labrador's commitment to international education," said the Honourable Darin King, Minister of Education. "The study also provides useful information for future consultations and decision-making among governments and stakeholders on attracting and retaining international graduates."

Working together and individually, the four Atlantic provinces are implementing strategic initiatives aimed at attracting and retaining more immigrants to the region to address demographic challenges associated with an aging population. The study highlights that they are ideal candidates for immigration as international students think highly of Atlantic Canada and are young, skilled, language proficient, and already integrated into local communities.

"We value the economic contribution of international students to our province," said Education Minister Marilyn More. "The province's new immigration strategy will identify recruitment of international graduates as a great way to tap into younger, well educated professionals who are already in Nova Scotia."

"International students provide a significant boost to Atlantic Canada's economy and generate millions in revenue," said the Honourable Donald Arseneault, Minister of Post-Secondary Education, Training and Labour.

The study underlines the cultural contributions of international students, strengthening and enhancing Atlantic Canada's profile and ties to the world. It also indicates that the market for international students is becoming more competitive since students make choices based on cost and institutional reputation. Since 2006, the price competitiveness of Atlantic Canadian universities has improved, and the number of international students in the region has increased steadily by approximately one-third, with 40 percent of students coming from China, the United States and India.

"Prince Edward Island's post-secondary institutions are world-class, and our cost of living and quality of education and life are attractive to international students," said the Honourable Allan Campbell, Minister of Innovation and Advanced Learning. "We are well positioned to attract more international students to our province and Atlantic Canada."
Source http://www.releases.gov.nl.ca/releases/2010/edu/0916n09.htm
Source http://www.releases.gov.nl.ca/releases/2010/edu/0916n09.htm

Annual Inflows of International Students

During 1998–2010, Atlantic Canada's educational institutions increased their share of annual inflows of international students in total Canadian inflows from about 4 % to about 6 %, the annual share increasing consistently over the period. In terms of actual numbers, annual inflows began rising at an accelerating pace even before 11 September 2001, as shown in Chart 1. During this period, national inflows of international students were also rising rapidly as the Canadian government (1) introduced expedited procedures for medical screening of international students, and (2) established Canada Education Centres in two major source countries of international students: People's Republic of China and Republic of Korea. The decline observed in actual inflows between 2003 and 2005 is consistent with the decline in national inflows.

International Students by Level of Study

Chart 2 shows the number of international students studying in Atlantic Canada by level of study. Most are university students, and their numbers are more than four times those of non-university students, whose numbers have, in turn, changed only slightly over the 12-year period. The population of international students grew

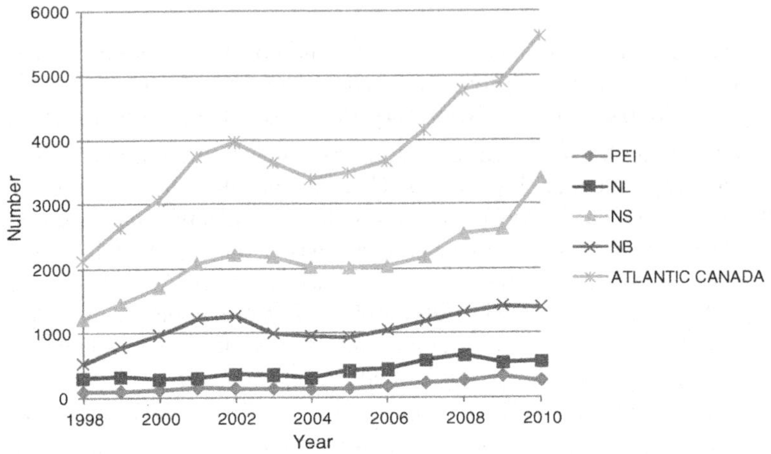

Chart 1 Total entries of international students into Atlantic Canada by province, 1998–2010. *Source* Table A23

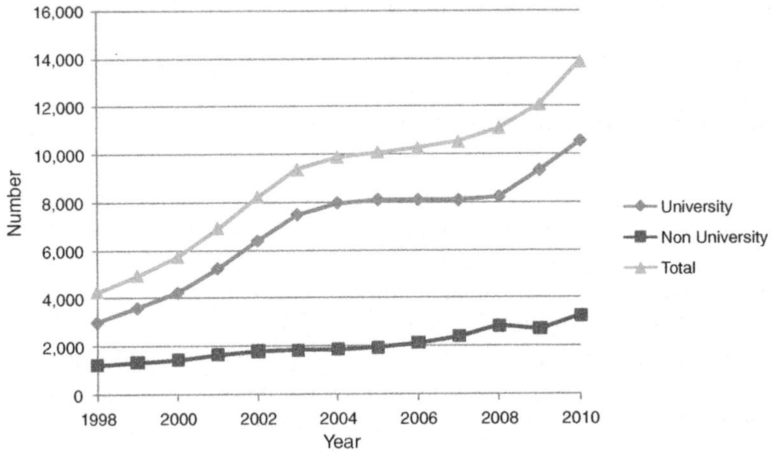

Chart 2 International students studying in Atlantic Canada by level of study, 1998–2010. *Source* Table A24

rapidly in Atlantic Canada over this period, as it did in Canada as a whole. That the number students in the region in a given year is greater than their inflow means that each international student takes more than 1 year to finish his/her studies, which is also true nationally. As suggested by the stock and inflow of students, an international student takes an average of about 3 years to finish his/her education in Atlantic Canada.

Source Countries of International Students

Chart 3 shows the top five source countries of international students. The presence of Saudi Arabia among these countries is worth noting since it is not included in the top five source countries of international students in Canada nationally. Atlantic universities individually promote their programs in the Middle East through collaboration with local universities and on-campus visits. For example, Memorial University of Newfoundland and Labrador especially aims to recruit students from Saudi Arabia and the Gulf States. In Nova Scotia, Mount Saint Vincent University's (MSVU) Internationalization Committee also connects with universities, colleges and private firms in Saudi Arabia interested in what exchanges and consultations MSVU could offer them. Saint Mary's University and Dalhousie University also recruit science and medical students there. Saudi Arabia hosts about 7 million expatriates, whose children will normally attend universities outside Saudi Arabia. The Tradewinds Flight Centre of Charlottetown, Prince Edward Island, extensively markets its pilot training in India and offers courses that suit the requirements of the Indian Director General, Civil Aviation. However, India is not among the top five source countries of students in the region. A recent pilot project launched by the Canadian government in India to expedite the processing of student applications is expected to increase significantly student applications from there to Canada.

The absence of Europeans in the top five source countries for students in Atlantic Canada also is worth noting, despite the region's proximity to Europe, while students from the United States are attracted to Canada mainly because of the lower cost of university education, usually lower value of the Canadian dollar and closeness to home.

Further, the establishment of Canadian Education Centres in the capital cities of China, India and Republic of Korea also has been instrumental in attracting rising numbers of students from those countries (Box 2).

Chart 3 Top five source countries of international students entering Atlantic Canada, 2010. *Source* Table A25

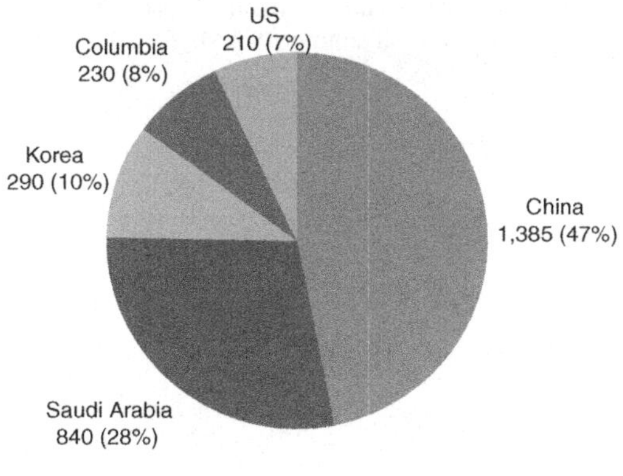

2 Majority of International Students Want to Live in Atlantic Canada After Finishing Their Education (Results of Another Survey)

In 2008, more than 8,000 international students were studying in universities across Atlantic Canada, representing 6 % of all international students in Canada, 1 % higher than it was in 2003. A survey of 135 international students in universities across Atlantic Canada was commissioned by ACOA in 2005. Survey results revealed that about 67 % of international students wanted to stay in Atlantic Canada after finishing their education (Lebrun and Rebelo 2006). This result was corroborated by another survey commissioned by the Maritime Provinces Higher Education Commission (MPHEC) in 2007, which found that 90 % of graduates of Maritime universities tend to stay in the Maritime provinces. The employment rate among the graduates of 2003 was 95 %, indicating that opportunities for the young and educated do exist in the region.

The ACOA survey also indicated that almost half of international students cited full employment, a welcoming community, social supports, cost of living and quality of life when considering applying for permanent residency, factors that Atlantic universities could include in their international marketing efforts.

References

Citizenship and Immigration Canada 2007, 2008 and 2011 (2007). *Facts and Figures*.

Hawthorne, L. (2005). Picking winners: The recent transformation of Australia's skill migration policy. *International Migration Review*, *39*(2), 663–696.

Lebrun, D., & Rebelo, S. (2006). The role of universities in the economic development of Atlantic Canada: A focus on immigration. Moncton, New Brunswick: ACOA-APECA Head Office.

Siddiq, F., Baroni, J., Lye, J., & Nethercote, W. (2010). The economic impact of post-secondary international students in Atlantic Canada. Report prepared for Council of Atlantic Ministers of Education and Training (CAMET). Halifax: Dalhousie University, School of Public Administration.

Chapter 6
Summary and Policy Recommendations

The 2011 census of Canada revealed a population growth of about 5.9 % from 2006 to 2011 (Statistics Canada 2012). However, Atlantic Canada's population grew only by about 1.87 % (from 2.28 million in 2006 to 2.32 million in 2011). Immigration and fewer losses in net-migratory exchanges with other provinces led to this positive growth, which had been close to zero in the previous five-year period. However, the rate of population growth was still the slowest in the country; as a result, Atlantic Canada's share of the national population declined from 7.2 to 6.9 % between 2006 and 2011. Also, due to declining fertility rates and aging in the post-World War II period, the natural growth of population in the region has been on a more rapid decline than it has been nationally. Out-migration of population has been a chronic problem faced by population planners in the post-World War II period, so provinces in Atlantic Canada are now relying more on immigration to meet their demographic challenges.

Immigration Trends in Atlantic Canada

Rough estimates indicate that international immigration over the 2001–2006 period contributed two-thirds of Canada's population growth. Over this period, an average of about 240,000 newcomers arrived in Canada each year for a total of some 1.2 million immigrants in 5 years. Only 2 % of them arrived in Atlantic Canada, however, preventing just one-sixth of the region's population decline.

Governments in Atlantic Canada view population decline and aging as resulting in adverse economic impacts for the region, which could include skill shortages, declining markets for goods and services, increased pressure on the future labour force to provide for programs for the elderly, and declining federal transfers. Population decline is also weakening the region's political representation in the federal government, which could aggravate the adverse impact of lack of population growth on the region's economy. As a result, provincial governments in the region have launched population strategies to reverse population decline and aging, one tool being to increase the region's share of international immigration,

S. A. H. Akbari, *Immigrants in Regional Labour Markets of Host Nations*, SpringerBriefs in Population Studies, DOI: 10.1007/978-94-007-6244-2_6,

a unified goal of all provincial governments. Various initiatives have been adopted to attract and retain immigrants in the region, and municipal governments, non-governmental organizations (NGOs), the private sector and the federal government have formed partnerships to implement them.

As a result of the above initiatives, immigration now plays an important role in shaping future population growth in Atlantic Canada. In fact, immigration in the region rose about 2.63 times during the 2000–2010 period (from 2,975 to 7,828). Arrivals increased in all four provinces during this period, with each also able to retain more immigrants than it had in earlier periods. However, immigration targets set by provincial governments are too low to achieve a fair share for the region. To maintain an even share in annual Canadian immigrant inflows relative to its population, the region should receive between 16,000 and 18,000 new immigrants each year. These numbers correspond to the region's share of the national population (7.2 %) and the national intake (0.7–0.8 % of population). Achieving this target may be difficult in the short term, but current initiatives are steps in the right direction that are producing desired results.

Just as in other regions of Canada, immigrants tend to settle in urban areas of Atlantic Canada, where about 67 % of the population lives. However, about 27 % of immigrants also live in rural Atlantic Canada. The top five source countries of these immigrants are the United States, the United Kingdom, Germany, the Netherlands and India. (In 2001, France was also included among these countries, while India was not.) Separate data on these immigrants' labour market performance are not available, but anecdotal evidence suggests that a large number of these immigrants may be involved in farming occupations.

Among recent immigrants, less than 20 % live in rural Atlantic Canada—mostly in moderate to highly rural areas—and their top five source countries are the United States, the United Kingdom, Germany, China and India. The large existing population of immigrants from the first three countries cited who are involved in farming occupations makes it possible that recent immigrants from those countries are as well. However, evidence also suggests that some engage in small business. Immigrants from China and India are more likely to be professionals, who came under the PNP to alleviate shortages of professionals, such as those of doctors in rural areas.

Immigrants in the Labour Market

Over the 1981–2010 period, proportionately larger groups of immigrants arrived in Atlantic Canada with higher educational attainments than non-immigrants. For example, during 2001–2010, more than half the immigrants aged 25 and older arrived with a university degree, while the latest data for non-immigrants reveal that only 15 % had a university degree in 2006 (the last census year). Immigrants who arrived in the late 1990s faced greater difficulties in labour markets, as reflected in their unemployment rates, which were higher than those not only of immigrants who had come in the past but also of non-immigrants. However, those who arrived more recently (during 2006–2010) did better in 2010. Their unemployment rate was comparable to that among non-immigrants, and the income gap between them and non-immigrants was the same as that between their counterparts and non-immigrants in 2006. The income gap between an average immigrant and non-immigrant closes completely 5 years after an immigrant's arrival, and the immigrant continues to earn more than a non-immigrant thereafter. Nationally, this gap closes after 15 years of an immigrant's stay in Canada. Overall, an immigrant earns a higher income than a non-immigrant in Atlantic Canada. However, those who obtained their post-secondary education in the Republic of Korea (South Korea), India and Pakistan face greater difficulties in labour market integration than other groups as their unemployment rates were higher.

To summarize, labour market outcomes of immigrants in Atlantic Canada are strong, except for those of some visible minority immigrants who obtained their education in the Republic of Korea, India, and Pakistan. Therefore, dissemination of more information about the educational systems of these countries to facilitate employers' hiring decisions is important. Canadian universities are experienced in assessing foreign credentials because they admit large numbers of international students, most of whom successfully complete their education. Therefore, these universities can be a useful resource in assessing an immigrant's education in a foreign country. Many Atlantic universities also have immigrant faculty members, whose knowledge can enhance the resourcefulness of those universities in assessing international credentials. Professional organizations and employers can also be involved in evaluating foreign education and experience.

One caveat in this analysis of unemployment rates is that no account is made for underemployment, which should be investigated in a separate study to further understand the issue of lack of foreign credential recognition in Atlantic Canada's labour markets.

Business immigration in the region has been traditionally low. The significant increase in the number of business immigrants during the mid-1990s was generated by specific situations, and most such immigrants live in Nova Scotia. Labour market employment and self- employment earnings of an average immigrant entrepreneur were the least among all classes of immigrants in 2008. Therefore, the Atlantic provinces need to achieve greater success in attracting business immigrants, who must have confidence in the local economy before being willing to invest in it. The experience of the1990s with business class immigrants, whose annual inflows rose and then fell rapidly due to the frustration of their expectations, suggests more government involvement is necessary in the settlement and integration of new business immigrants and their families. Assistance should also be provided in developing contacts with the existing network of entrepreneurs. The "Entrepreneur Forum" introduced by the province of Nova Scotia is a step in the right direction. Immigrant entrepreneurs should also be included in the promotional programs to market the region to potential immigrant entrepreneurs in foreign countries. To avoid any possible abuse of the system, however, close government monitoring of the local connections established by new arrivals is essential. Partnerships with commercial banks also should be sought to facilitate the availability of financial credit to new arrivals under business class.

The region is now receiving more highly skilled immigrants than in the past. During 2000–2004, the region also achieved a net in-migration of skilled immigrants from other provinces, which helped increase the overall retention rate of those who arrived during that period (Akbari 2009). Their numbers had declined between 1995 and 2003, the effect of which was quite broad based among professionals in the region. Since the adoption of certain initiatives, such as the PNP and strategies to fast track the integration of international medical graduates, however, the numbers of highly skilled immigrants have started to rise again. These programs have also been the main source of attraction for skilled immigrants from provinces outside Atlantic Canada, most of whom are working in rural areas through special arrangements. Data on the occupational distribution of these immigrants and their economic performance should be analyzed to understand how well they have integrated into the region.

Finally, it is also to be noted that in this study, the comparative analysis of labour market performance of immigrants and non-immigrants did not control for differences in certain demographic and human capital characteristics. Such an analysis should be conducted in future within a multi-variate framework and will require access to larger sample of census, perhaps through Statistics Canada's research data centers.[1]

International Students

International students in Canada offer a large pool of potentially highly skilled immigrants. On average, an international student takes about 3 years to finish his/her education in Atlantic Canada. Since most are university students (four times the number of non-university students), conversion of their residency status into that of landed immigrant would increase the supply of highly skilled immigrants who do not face the same labour market challenges as those who acquired their education at a non-Canadian university. They can also integrate more easily into Canadian society because of their proficiency in English or French (or both) and knowledge of Canadian culture.

In 2010, close to 14,000 international students were studying in universities across Atlantic Canada. To attract more international students, universities could focus more on students from countries with existing communities in Atlantic Canada. A large immigrant population from the United States and the United Kingdom lives in Atlantic Canada, and these countries' proximity could make it

[1] However, controls for age and an immigrant's length of stay in Canada were introduced when comparing the earnings of immigrants and native-born (In Chap. 4 Charts 5 and 6), although not in a multivariate framework.

easier to attract more students from there. Other factors that should be strongly promoted among students in these two countries about university education in Atlantic Canada could include the competitive tuition fees and high quality of education. Another dominant immigrant community in the region is Middle Eastern, whose members could help attract students from their countries of former residence. Also, some Middle Eastern countries, including Saudi Arabia, Kuwait and the United Arab Emirates, do not permit children of expatriates to attend local universities. Atlantic Canadian universities could also target those students.

Another way for universities in Atlantic Canada to attract international students could be to offer more professional degree programs at graduate level. There is a rising demand for professional degrees around the world as was suggested by Canadian Association of Graduate Studies (2006, p. 31): "Probably the most important trends affecting the evolution of master's degrees in Canada and abroad are growth in professional degrees and enrolment." To be internationally competitive, universities in Atlantic Canada should offer more professional degrees in diversified disciplines. A study could identify the major disciplines that are attractive for international students seeking a professional graduate degree.

Today, most international students, as well as immigrants, come from China. As this population grows, it is expected that more Chinese students will also be attracted to the Atlantic region. Therefore, strengthening English as a Second Language (ESL) programs and industry-specific language training as part of university curricula will also attract students from China and other non-English-speaking countries.

At present, from the time of entry into Canada, it takes about ten years for an international student to acquire landed immigrant status. As much as possible, the federal government could help facilitate the landing process for international students and procedures for obtaining a work permit after graduation. More provincial governments also could allow universities to participate in the PNP in collaboration with private employers.

Some Policy Implications that Emerge from Statistical Findings of Present Study

With increased immigration, issues relating to the social and economic impacts of immigration that are often raised in the public circles of the immigrant-abundant regions of western and central Canada could occupy a central place in public policy discussions in Atlantic Canada as well. Some issues include the impact of immigrants on the public treasury, poverty, employment and the wages of the native born. The present study has provided some broad information that can also be useful to public policy discussions around these issues.

Several policy implications emerge from the findings reported in this study: (1) rural areas can be developed as immigrant destinations; (2) farming opportunities and the presence of immigrant communities in Atlantic Canada should be promoted in the United States, Europe and Asia to attract more immigrants from there. Promotional materials being designed under the federal-provincial Atlantic Population Table's initiative to attract immigrants to the region should also promote farming opportunities and an existing immigrant population to potential immigrants from the United States and Europe; (3) job availability for professionals should be promoted in all countries around the world by referring to the shortages of skilled labour; and (4) assistance with immigrant settlement in the form of language programs and cultural events should be a priority, especially for immigrants from countries other than the United States and those in Europe. Also, investigating the motivation for choosing a rural destination can provide information that could be used to attract future immigrants.

During the 1920s and 1930s, the Canadian government succeeded in using immigration in its rural population strategy applied to western Canada and the Prairie provinces, with the objective of developing agricultural land. Given Atlantic Canada's relative proximity to Europe and the United States, a similar strategy for rural re-population in Atlantic Canada should be considered by regional and federal policy makers for this region. The formation of a partnership between the Nova Scotia Office of Immigration and the Department of Agriculture to attract immigrant farmers to rural Nova Scotia through the Provincial Nominee Program represents a beginning in this regard.

Diversification of economies in Atlantic Canada is an important goal of federal and provincial governments. While resource-based industries continue to be important, there also has been diversification into new value-added products and services. Recent examples include the agreement between the governments of Newfoundland and Labrador and of Nova Scotia on the Churchill Falls electricity generation project and the award of a major shipbuilding project by the federal government to Irving Shipbuilding in Halifax. Projects like these will generate demand for new skilled and unskilled workers, which are likely to slow youth outmigration rates and increase the inflow of workers from other parts of Canada, as well as from abroad given the changes taking place in the age composition of population due to declining fertility rates. The analysis presented in this book has shown that immigration can be used to meet labour market shortages because immigrants will stay in the region if economic opportunities are available to them. Hence, governments and the private sector need to determine how much of the region's future labour demand can be met from domestic sources and how much will need to be met through immigration.

References

Akbari, A. H. (2009). Socioeconomic and Demographic Profiles of Immigrants in Atlantic Canada Report prepared for Atlantic Canada Opportunities Agency (www.atlantic.metropolis.net).

Canadian Association of Graduate Studies. (2006). A profile of master's degree education in Canada (http://www.cags.ca/documents/publications/CAGS-Master.pdf).

Statistics Canada. (2012). Population and dwelling counts, for Canada, provinces and territories, 2011 and 2006 censuses (table). Population and dwelling count highlight tables. 2011 Census. Statistics Canada Catalogue no. 98-310-XWE2011002. Released February 8, 2012 http://www12.statcan.gc.ca/census-recensement/2011/dp-pd/hlt-fst/pd-pl/File.cfm?T=101&SR=1&RPP=25&PR=0&CMA=0&S=50&O=A&LANG=Eng&OFT=CSV (accessed March 10, 2012).

Epilogue

Due to their slow or declining population growth rates, many developed countries in the western world are relying more and more on immigration to maintain their labour supply, especially in professional and skilled jobs. In Canada, according to a 2011 government report (Citizenship and Immigration Canada 2011a), immigration is projected to account for all net labour force growth within the next decade and all population growth within the next two decades. It is logical, therefore, that The Migration Policy Institute (2005) considers growing competition for skilled immigrants among the top issues of migration in the developed world.

Smaller areas in developed countries are affected more by declining population as their working age population migrates to larger urban areas. Therefore, realizing the economic, social and political impacts of population decline and aging, these smaller areas are now playing a greater role in their national immigration programs and are expected to continue doing so in future. While studies that use national data show deterioration in the performance of immigrants in Canada, statistical facts from the Atlantic region of Canada demonstrate that immigration policy can be effective in increasing economic returns from immigration by encouraging the settlement of immigrants in smaller areas.

S. A. H. Akbari, *Immigrants in Regional Labour Markets of Host Nations*, SpringerBriefs in Population Studies, DOI: 10.1007/978-94-007-6244-2,

Appendix

Table A1 Immigrant arrival rates (per thousand)

Province	2001	2006	2010
BC	9.85	10.23	9.75
AL	5.51	6.3	8.77
SK	1.74	4.77	7.28
MN	4.1	8.75	12.79
ON	13.03	10.35	8.94
QC	5.2	5.92	6.83
NB	1.09	2.25	2.83
NS	1.87	2.83	2.55
PE	1	4.15	18.14
NL	0.76	1.01	1.34
CA	8.35	7.96	8.23

Source Immigrant inflows are based on special tabulations obtained from Citizenship and Immigration (2011b). Provincial population data are obtained from Statistics Canada (2011, special tabulations)

Table A2 Components of population growth rates in Atlantic Canada in the post-world war II period

Component	Late 1940s	1950s	1960s	1970s	1980s	1990s	2000–2005	2006–2010
Rate of growth	4.39	15.4	8.62	10.29	4	−0.18	−0.44	0.62
Due to natural growth	9.32	23.18	17.52	10.78	7.11	3.77	0.56	0.52
Due to net migration	−4.93	−7.78	−8.9	−0.49	−3.11	−3.95	−1	0.1

Source Table A3

S. A. H. Akbari, *Immigrants in Regional Labour Markets of Host Nations*, SpringerBriefs in Population Studies, DOI: 10.1007/978-94-007-6244-2,

Table A3 Birth and death rates in Atlantic Canada, 1945–2010 (per 000)

Year	Population ('000)	Birth Rate	Death Rate
1945	321.576	34.9	10.4
1946	1,509.671	32.5	10.1
1947	1,534.227	34.3	9.9
1948	1,560.201	31.8	9.6
1949	1,575.972	31.4	9.3
1950	1,597.040	31.1	9.4
1951	1,617.869	29.4	9.0
1952	1,652.839	30.2	8.5
1953	1,680.144	29.9	8.4
1954	1,708.595	30.4	8.1
1955	1,736.529	30.6	8.3
1956	1,764.057	30.0	8.2
1957	1,786.238	30.4	8.2
1958	1,811.924	29.1	8.1
1959	1,843.250	28.8	8.3
1960	1,866.581	28.6	7.9
1961	1,896.714	28.7	7.8
1962	1,925.826	27.9	8.0
1963	1,944.636	27.3	7.9
1964	1,957.895	26.1	7.7
1965	1,968.079	24.4	7.8
1966	1,975.475	22.4	7.8
1967	1,988.002	20.9	7.9
1968	2,008.064	20.1	7.8
1969	2,028.007	19.9	7.7
1970	2,036.049	19.7	7.8
1971	2,083.21	19.8	7.6
1972	2,103.608	19.1	7.7
1973	2,129.287	18.1	7.7
1974	2,149.061	17.6	7.7
1975	2,177.777	17.5	7.5
1976	2,205.947	17.1	7.5
1977	2,221.121	16.6	7.4
1978	2,233.465	16.0	7.2
1979	2,245.514	15.7	7.2
1980	2,255.372	15.6	7.4
1981	2,260.162	15.3	7.2
1982	2,263.878	15.0	7.3
1983	2,287.397	14.8	7.3
1984	2,304.587	14.4	7.3
1985	2,316.029	14.3	7.4
1986	2,318.848	13.9	7.5
1987	2,325.257	13.5	7.4

(continued)

Table A3 (continued)

Year	Population ('000)	Birth Rate	Death Rate
1988	2,331.836	13.4	7.5
1989	2,345.674	13.6	7.6
1990	2,358.379	13.7	7.6
1991	2,370.549	12.9	7.5
1992	2,378.508	12.6	7.6
1993	2,384.891	12.1	7.7
1994	2,384.959	11.8	7.9
1995	2,380.875	11.3	7.9
1996	2,379.03	11.0	7.9
1997	2,371.919	10.5	8.2
1998	2,358.013	10.2	8.4
1999	2,353.995	10.1	8.1
2000	2,348.774	9.7	8.3
2001	2,340.951	9.5	8.2
2002	2,340.753	9.3	8.3
2003	2,342.621	9.3	8.4
2004	2,343.866	9.2	8.5
2005	2,338.319	9.1	8.6
2006	2,331.917	9.2	8.5
2007	2,325.849	9.4	8.7
2008	2,329.377	9.4	8.9
2009	2,337.688	9.3	9.1
2010	2,346.266	9.3	9.3

Source Column 2 is based on Statistics Canada (2011, special tabulation)
Columns 3 and 4 are based on Statistics Canada, CANSIM Table 530001 (Accessed through CHASS, July 11, 2011)
Pre-1949 data for NL were provided by the NL office of immigration and multiculturalism

Table A4 Provincial representations in the Canadian house of commons based on current rules and projected population growth, 1976–2021

Province	House Seats 1976–2000	House Seats 2001–2011	House Seats 2012–2021
NL	2.51	2.30	2.26
PE	1.43	1.31	1.29
NS	3.94	3.61	3.55
NB	3.58	3.28	3.23
QC	26.88	24.59	24.19
ON	34.05	34.75	35.48
MN	5.02	4.59	4.52
SK	5.02	4.59	4.52
AL	7.53	9.18	9.35
BC	10.04	11.80	11.61

Source Based on Tomlin (2007)

Table A5 Immigrants (principal applicants and dependents) destined for each Atlantic province during 1981–2010

Year	Provinces				
	NS	NB	NL	PEI	Total
1981	1403	988	480	126	4372
1982	1254	751	407	165	3987
1983	833	554	275	105	3370
1984	1035	600	299	109	3619
1985	972	607	323	113	3564
1986	1094	640	274	168	3720
1987	1223	641	455	160	3851
1988	1298	674	410	152	3960
1989	1473	902	466	158	4364
1990	1571	853	552	176	4414
1991	1499	686	636	150	4176
1992	2360	757	788	151	5109
1993	3022	706	804	163	5721
1994	3469	627	565	161	6090
1995	3579	643	573	161	6217
1996	3224	717	581	150	5937
1997	2833	663	414	144	5493
1998	2042	723	402	136	4763
1999	1595	660	424	135	4254
2000	1610	759	417	189	4369
2001	1699	798	392	134	4498
2002	1419	705	407	107	4126
2003	1474	665	359	153	4142
2004	1770	795	579	310	4569
2005	1929	1091	496	330	5025
2006	2586	1646	508	565	6238
2007	2523	1643	546	992	6173
2008	2651	1856	616	1443	6566
2009	2424	1913	603	1723	6663
2010	2408	2125	714	2581	7828

Source Data from 1981–2005 are from Permanent Resident Data System (PRDS), micro-data, CIC. The 2006–2010 data are from Citizenship and Immigration Canada (2011b), (www.cic.gc.ca; accessed August 11, 2011)

Table A6 Immigration in Atlantic Canada by category, 2000–2010

Immigration category	2000	2001	2002	2003	2004	2005	2006	2007	2008	2009	2010
Family class	595	755	635	705	900	835	915	855	1,040	920	825
Economic Immigrants	1,600	1,565	1,325	1,220	1,870	2,335	3,660	4,055	4,770	5,075	6,310
Refugees	780	705	560	505	565	600	600	610	570	520	590
Other immigrants	–	–	120	215	115	80	135	185	190	150	110
TOTAL	2,970	3,025	2,640	2,655	3,455	3,850	5,305	5,700	6,565	6,660	7,830

Source Special tabulations based on Citizenship and Immigration Canada (2011a)

Table A7 Annual arrivals of provincial nominees in Atlantic Canada by province, 2000–2010

Province	2000	2001	2002	2003	2004	2005	2006	2007	2008	2009	2010
Principal applicants and dependents											
NB	25	70	105	150	165	440	970	920	1,035	1,170	1,350
NL	0	35	40	40	175	85	80	70	100	180	215
NS	0	10	0	0	65	330	860	900	870	830	635
PEI	0	0	10	45	145	200	425	815	1,250	1,530	2,420
Total	25	115	155	230	535	1,055	2,330	2,705	3,250	3,715	4,630
Principal applicants											
NB	10	30	45	60	65	150	325	325	345	395	460
NL	0	10	10	15	55	30	30	25	60	85	125
NS	0	–	0	0	20	105	255	290	315	300	295
PEI	0	0	–	15	45	65	140	260	385	490	770
Total	10	50	60	90	180	345	750	890	1,105	1,265	1,640

Source Special tabulations based on Citizenship and Immigration Canada (2011a)

Table A8 Annual share of provincial nominees in immigration, Atlantic Canada by province, (%)

Province	2005	2006	2007	2008	2009	2010
NB	40.33	58.93	56.00	55.77	61.16	63.53
NL	17.14	15.75	12.82	16.23	29.85	30.11
NS	17.11	33.26	35.67	32.82	34.24	26.37
PE	60.61	75.22	82.16	86.63	88.80	93.76
AC-total	27.43	43.92	47.42	49.50	55.76	59.15

Source Special tabulations based on Citizenship and Immigration Canada (2011a)

Table A9 Population distribution in Atlantic Canada, 2001–2006

	Total population	Non-immigrants	Immigrants	Recent immigrants	Non-permanent resident
CMA/CA (Urban)	67.18	66.78	72.81	81.74	79.33
Total MIZ (Rural)	32.82	33.22	27.19	18.26	20.67
Strong MIZ	4.02	4.10	2.97	1.91	0.80
Moderate MIZ	13.47	13.65	11.01	6.41	8.54
Weak MIZ	14.79	14.93	12.71	9.45	11.01
No MIZ	0.54	0.54	0.50	0.50	0.32

Source Statistics Canada (rural secretariat community information database, bo-0355, Table 3—profile of statistical area classification, 2006 census, 20 % sample; B20/20 files)

Table A10 Immigrants in weak metropolitan influence zones of Atlantic Canada by source country, 2006

Rank	Country	Immigrants		Rank	Country	Recent immigrants	
1	USA	4745	39.15	1	USA	470	33.81
2	UK	2875	23.72	2	UK	170	12.23
3	Germany	960	7.92	3	Germany	140	10.07
4	Netherlands	765	6.31	4	China	65	4.68
5	India	160	1.32	5	India	35	2.52
6	All others	2615	21.58	6	All others	510	36.69

Source Statistics Canada (rural secretariat community information database, bo-0355, Table 3—profile of statistical area classification, 2006 census, 20 % sample; B20/20 files)

Table A11 Age distributions of recent immigrants and non-immigrants, Atlantic Canada, 1986–2006

Age group	1986		1991		1996		2001		2006	
	Recent immigrants	Non-immigrants	Recent immigrants	Non-immigrants	Recent immigrants	Non-immigrants	Recent immigrants	Non-immigrants	Recent immigrants	Non-immigrants
15–24	19.53	24.13	19.46	20.43	11.22	18.53	10.12	17.02	10.88	15.86
25–54	52.06	40.40	58.86	42.19	62.69	40.57	68.38	36.57	66.44	33.54
45–64	16.53	22.19	13.24	23.27	20.75	26.91	18.35	31.28	19.55	33.88
65+	11.89	13.28	8.43	14.11	5.33	13.99	3.15	15.13	3.13	16.73
Total	100.00	100.00	100.00	100.00	100.00	100.00	100.00	100.00	100.00	100.00

Source Akbari (2009)

Table A12 Immigrant retention rates for Atlantic Canada, 1986–2006

Period	New immigrants arriving[a]	New immigrants residing at end of period	Retention rate (%)
	(1)	(2)	(3) =(2)/(1)×100
1981–1986	12212	9200	75
1986–1991	14359	10200	71
1991–1996	23136	11005	48
1996–1901	19031	9940	52
2001–2006	20920	13490	77

[a] Because censuses are conducted in mid-June, new immigrants residing in the region in the last year of each period do not include those who arrived in the latter half of those years, i.e. 1986, 1991, 1996 and 2001. Hence, arrival data for only the first five months of each of these years were used

Source Data in column (1) for the period 1981–2001 are based on permanent resident data system (PRDS - microdata); for 2001–2006 are based on special tabulations obtained from CIC. Column (2) data are based on Canadian population censuses obtained from the following sources: (1) For 2001–2006 data, catalogue no. 97-557-XCB2006019, (2); for 1996–2001 data, Statistics Canada, catalogue number 97F0009XCB-2001004. (3) for 1991–1996 data, Statistics Canada catalogue number 93F0023XDB96003; (4) for 1986–1991 data, Citizenship and Immigration (2002, selected charts based on census 1991); (5) for 1981–1986 data, Statistics Canada population census 1986, public use microdata

Table A13 Recent immigrants destined for Atlantic Canada and non-immigrants by education levels, 1986–2010

Year	High school or less		University degree	
	Immigrants	Non-immigrants	Immigrants	Non-immigrants
1986	45.43	61.44	26.56	7.58
1991	40.37	56.74	30.76	8.83
1996	35.30	51.23	39.61	10.98
2001	30.63	47.84	44.83	12.47
2006	24.78	41.13	51.04	15.30
2010	21.8	41.0	48.3	16.6

Note Recent immigrants are those who arrived within 5 years of the census / survey year

Source For the 1986–2006 period: (1) Permanent resident data system (PRDS—microdata, CIC) for immigrants. Variables used: "prov", "ed_qua", "fage". High school or less education = None + Secondary or less. (2) The Canadian population censuses (PUMF, 1986–2001, individual files) for non-immigrants until 2001. Variables used: "province or territories", "immigrant status indicator", "age", "highest level of schooling". For the 1986 census, "place of birth" is used instead of "immigrant status indicator" because the latter is not available to determine whether the individual is a non-immigrant. High school or less education = less than grade 5 + grades 5–8 + grades 9–13 + secondary school graduation certificate. For the 2006 census-based data, Statistics Canada catalogue no. 97-560-XCB2006025. Immigrant data for 2010 are based on special tabulations obtained from Citizenship and Immigration (2011a). Non-immigrant data for that year are based on LFS (special tabulations obtained from Statistics Canada). All data are reported for individuals aged 25 and over

Table A14 Labour force net growth rate and contribution of recent immigrants to the growth of the labour force, Atlantic Canada, 1981–2001

Period	Total growth of labour force	Growth owed to new immigrants	Growth without new immigrants	Immigrants' contribution to labour force growth (%)
	(1)	(2)	(3)	(4) =(2)/(3)×100
1981–1986	83320	3790	79530	4.77
1986–1991	88610	4795	83815	5.72
1991–1996	−20850	4975	−25825	19.26
1996–2001	13630	4245	9385	45.23
2006–2010	36000	9100	26900	34.0

Source Until 2001, calculations based on population census data provided in Statistics Canada catalogue 97F0012XCB2001003, accessed May 5, 2007, on web site: www.statcan.ca. Comparable data from 2006 census were not available. 2006–2010 data are based on the labour force survey (special tabulations obtained from Statistics Canada)

Table A15 Atlantic Canada labour market statistics: participation rate, unemployment rate, employment income, and government transfer payments as a percentage of total income for immigrants and non-immigrants, 1981–2010

Census / Survey year	1981	1986	1991	1996	2001	2006	2010
Labour Force Participation Rate							
Non-immigrants	60.05	61.89	64.24	61.76	62.53	62.8	63.6
Immigrants	59.00	60.98	62.12	59.35	58.41	58.6	58.2
Recent immigrants	64.21	63.02	70.05	65.55	64.32	64.4	67.4
Unemployment Rate (%)							
Non-immigrants	16.46	20.04	19.80	19.09	15.75	11.7	10.6
Immigrants	10.09	11.98	12.17	11.34	9.38	7.8	7.7
Recent immigrants	12.17	15.92	17.52	18.69	18.37	11.5	14.3
Employment Income							
Immigrant relative to non-immigrant	1.39	1.34	1.36	1.18	1.21	1.25	1.10
Recent immigrant relative to non-immigrant	1.10	0.76	N/A	1.00	0.90	0.88	0.90
Government Transfers (% of total income)							
Non-immigrants	14.48	18.22	18.36	20.99	18.09	17.63	NA
Immigrants	9.25	11.55	12.61	17.55	15.90	14.55	NA
Recent immigrants 1	2.99	5.87	N/A	12.41	9.40	10.78	NA

Sources and notes 1. For 1981–2001, labour force participation and unemployment rates are based on "Historical Labour Force Activity (Based on the 1971 Concepts) (8), Immigrant Status and Period of Immigration (10), Age Groups (18), Marital Status (7) and Sex (3) for Population 15 Years and Over, for Canada, Provinces and Territories, 1971, 1981–2001 Censuses—20% Sample Data". Ottawa: Statistics Canada, March 25, 2003, Census of Canada, Catalogue number 97F0012XCB2001003. The rest of the data are based on special tabulations performed by the author based on population censuses (PUMF, 1981–2001, individual files). 2. All income data are averages. Census-based data on recent immigrants include those who arrived within five years of the census. Because they include those arriving in the census year and in the year prior to it, they do not reflect their entire year's performance. Employment income is equal to the sum of wages and salaries and self-employment income and is reported only for those who were employed. 3. The following variables were used to obtain government transfers as a percentage of total income: For the 1991, 1996 and 2001 censuses, "total income" and "government transfer payment" were available directly. For the 1986 census, "total income", "Old age sec pension and guaranteed income supplement (OASGIP)", "Canada, Quebec pension plan benefits (CQPPBP)", "Family allowances (FAMALP)", "Federal child tax credits (CHDCRP), "Unemployment insurance benefits (UICBNP)" and "Other government transfer payments (GOVTIP)" were added. For the 1981 census, "total income", "OAS, GIS, CQPP", "Unemployment insurance benefits (UICBN)" and "Other government transfer payments (GOVTI)" were added. 4. All labour force and income information from the census are for the year prior to the census year. 5. The 1991 PUMF does not provide separate data on recent arrivals in the Atlantic provinces. 6. The 2006 census-based data are from… (something missing here?) 7. Data for 2010 are based on 2010 LFS (special tabulations obtained from Statistics Canada). Income data for that year are the weekly wages. Recent immigrants in 2010 survey are those who arrived during 2006–2010

Table A16 Labor market indicators, immigrants and non-immigrants in Canada and Atlantic Canada, 2010

	Canada			Atlantic Canada		
	All immigrants	Recent immigrants	Non-immigrants	All immigrants	Recent immigrants	Non-immigrants
Unemployment rate (%)	9.8	15.8	7.5	7.7	14.3	10.6
Labor force participation rate (%)	62.4	67.3	68.4	58.2	67.4	63.6
Weekly wage ($)	798	641	827	811	685	734

Source Statistics Canada, labour force survey, 2010, special tabulations

Table A17 Age wage Profile of an immigrant and a non-immigrant resident of Atlantic Canada, 2010

Age groups	Non-immigrants		Immigrant	
	Average weekly wages ($)	No. of employees ('000)	Average weekly wages ($)	No. of employees ('000)
15+	734.4	916.40	811.54	32.20
15–24	351.09	145.20	283.01	2.60
25–54	820.38	625.50	824.99	21.20
55+	747.23	145.70	940.51	8.40
15–64	737.74	901.80	812.37	30.80

Source Statistics Canada, labour force survey, 2010, special tabulations

Table A18 Average weekly earnings of an immigrant by length of stay and non-immigrant, Canada, 2010

Year	Canada		Atlantic Canada	
	Earnings ($)	No. of employees	Earnings ($)	No. of employees
Total	820.65	14,371.20	737.32	958.8
Total landed immigrants	797.83	2,832.30	811.54	32.20
With 5 years or less stay	641.27	433.10	684.56	6.70
With 5–10 years stay	729.98	467.20	780.79	4.10
With 10+ years stay	849.34	1,932.00	856.93	21.40
Non-immigrants	827.44	11,295.90	734.40	916.40

Source Statistics Canada, labour force survey, 2010, special tabulations

Table A19 Immigrants destined for Atlantic Canada's labour force by skill classification[a], 1981–2010

Year	Highly skilled	Low skilled	Medium skilled
1981	454	294	445
1982	401	238	377
1983	289	144	214
1984	255	223	259
1985	283	227	241
1986	330	236	268
1987	400	253	360
1988	389	290	388
1989	447	343	491
1990	428	307	414
1991	397	206	277
1992	444	199	282
1993	487	186	366
1994	522	142	297
1995	664	155	339
1996	645	139	322
1997	589	130	279
1998	503	73	221
1999	519	88	183
2000	520	90	185
2001	550	65	200
2002	490	50	150
2003	465	60	145
2004	620	65	165
2005	755	70	160
2006	1120	80	240
2007	1240	130	330
2008	1495	150	370
2009	1480	170	365
2010	1820	275	360

Source data for 1981–1997 are from Akbari (2009). Data for 1998–1999 and for 2000–2010 are based on special tabulations obtained from Citizenship and Immigration (2007 and 2011a)

[a] National Occupational Classifications (NOC) were further classified as highly skilled = "O" and "A"; medium skilled = "B"; low skilled = "C" and "D". Detailed definitions of NOC are provided in Table A5. *Source* PRDS – microdata, Citizenship and Immigration Canada. Variables used: "NOC2", "PROV"

Table A20 Provincial distribution of highly skilled immigrants in Atlantic Canada, 2001, 2006 and 2010

Province	2001		2006		2010	
	Managers	Professionals	Managers	Professionals	Managers[a]	Professionals
NL	333	2261	680	1910	500	1,800
PE	374	448	245	520	300	800
NS	3836	6823	3190	6695	3100	7000
NB	1478	3437	1555	3050	1800	3800

Source Statistics Canada, census 2001, 2006 target group profile, and LFS 2010, special tabulations
[a] Includes all those who were listed in management occupations by LFS

Table A21 Labour Market Earnings of Immigrants by Class, 2008

Class	Arrival period: 1993–2007		Arrival period: 2003–07	
	Number	Income	Number	Income
Family	3,305	31,824	1,500	28,536
Business (PA)	190	30,725	40	15,195
Skilled worker (PA)	2,700	57,793	1,155	49,935
Other economic	2,675	24,717	790	20,256
Refugee	2,730	24,028	1,975	22,246
Other	755	30,303	395	26,653

Source Summary Tables based on 2008 IMDB

Table A22 Employment income[a] reported by tax-filers and their numbers, Atlantic provinces, 2008 taxation year

Province	No. of tax-filers reporting employment income	Total income ($ '000)	Average income ($)
NS	460,020	15,116,117	32,860
NB	393,910	12,412,216	31,510
PE	70,430	2,019,906	28,679
NL	271,130	8,424,108	31,070
Atlantic Canada	1,195,490	37,972,347	31,763

Source Canada revenue agency, income statistics 2010–2008 taxation year, basic Table 3 (for each province lines 4 and 6). [a] Includes "employment income" and "other employment incomes"

Table A23 Total entries of international students in Atlantic Canada by province and in Canada, 1998–2010

Province	Year												
	1998	1999	2000	2001	2002	2003	2004	2005	2006	2007	2008	2009	2010
PE	85	95	110	140	135	130	125	135	165	215	255	325	255
NL	305	315	285	295	365	350	295	410	430	575	650	530	550
NS	1,215	1,455	1,705	2,075	2,205	2,175	2,015	2,005	2,025	2,170	2,530	2,605	3,395
NB	525	770	965	1,225	1,255	985	955	935	1,045	1,185	1,325	1,420	1,400
Atlantic Canada total	2130	2635	3065	3735	3960	3640	3390	3485	3665	4145	4760	4880	5600
CANADA total	48,025	58,410	69,075	80,905	76,930	69,705	66,115	67,880	71,790	74,035	79,525	85,180	96,160

Source Special tabulations based on citizenship and immigration Canada (2011a)

Table A24 International students studying in Atlantic Canada by Level of Study, 1998–2008

	Year												
	1998	1999	2000	2001	2002	2003	2004	2005	2006	2007	2008	2009	2010
University	3,020	3,615	4,250	5,270	6,415	7,500	7,985	8,120	8,105	8,095	8,225	9,325	10,565
Non-University	1,255	1,340	1,480	1,660	1,815	1,855	1,885	1,965	2,145	2,420	2,860	2,735	3,285
Total	4,275	4,955	5,730	6,930	8,230	9,355	9,870	10,085	10,250	10,515	11,085	12,060	13,850

Source Special tabulations based on Citizenship and Immigration Canada (2011a)

Table A25 International students studying in Atlantic Canada and Canada by top five source countries, as of December 1, 2008

Source country	Atlantic Canada	Canada
China	3,395	42,154
United States of America	910	27,440
Korea, Republic of	707	11,317
Saudi Arabia	595	X
India	361	7,314
France	X	8,353

Source Special tabulations based on Citizenship and Immigration Canada (2008)

References

Akbari, A. H. (2009). *Socioeconomic and demographic profiles of immigrants in Atlantic Canada report prepared for Atlantic Canada opportunities agency.* www.atlantic.metropolis.net.

Citizenship and Immigration Canada (2002). *Recent immigrants in the Halifax metropolitan area* (October, Strategic Research and Review, Research Unit-Ottawa).

Citizenship and Immigration Canada (2007). *Facts and figures.* Research and Evaluation Branch.

Citizenship and Immigration Canada (2008). *Facts and figures.* Research and Evaluation Branch.

Citizenship and Immigration Canada (2011a). *Annual report to parliament on immigration.*

Citizenship and Immigration Canada (2011b). *Facts and Figures.* Research and Evaluation Branch.

Hawthorne, L. (2005). Picking winners: the recent transformation of Australia's skill migration policy. *International Migration Review, 39(2),* 663–692.

Statistics Canada (2011) *Estimates of total population, Canada, provinces and territories* (Special tabulations, Demography Division).

Tomlin, B. (2007). *The seat shortage: changing demographics and representation in the of 266 commons* C.D. Howe Institute (e-brief, May 29, 2007).

Made in the USA
Las Vegas, NV
13 November 2024

11643073R00063